CRITICAL THINKING & LOGICAL REASONING WORKBOOK-10

10

GIFT OF LOGIC™ SERIES

An Essential Resource for Everyone

Boost Your Thinking Skills

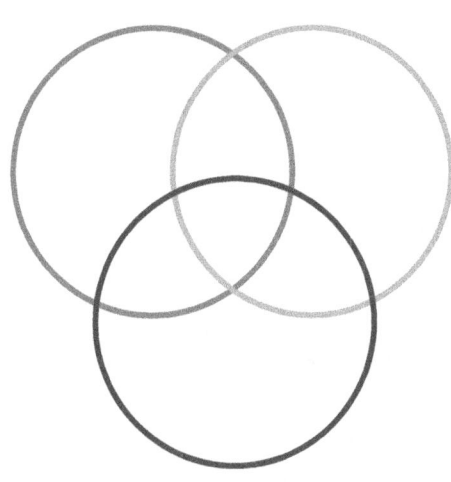

Verbal Reasoning
Analytical Reasoning
Pictorial Reasoning

THIRD EDITION

| FOR GRADES 6-12 | STUDENTS, TEACHERS, AND PARENTS |

Ranga Raghuram

GIFT OF LOGIC™

Gift Of Logic, Inc

http://www.giftoflogic.com
sales@giftoflogic.com

Critical Thinking and Logical Reasoning Workbook-10
ISBN-13: 978-1494833169
ISBN-10: 1494833166

Third Edition
1-2014

Copyright © 2009 Gift Of Logic, Inc. All rights reserved. No part of this publication may be reproduced, stored in a retrieval system, transmitted in any form or by any means, electronic, mechanical, photocopying, recording or otherwise, without the written permission of the publisher.

License: This book is licensed for use by one person only. Use of this book in a group setting (classroom, workshop, etc) without the written permission of the publisher is prohibited. Unauthorized duplication is strictly prohibited by law. Contact the publisher at sales@giftoflogic.com for classroom/school/group licensing.

GIFT OF LOGIC™
CRITICAL THINKING & LOGICAL REASONING CURRICULUM
12 WORKBOOKS TO BOOST YOUR THINKING SKILLS

For Kindergarten, Grade 1, and Grade 2

Workbook # 0 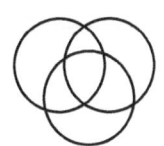		
	Verbal Reasoning	Finding the truth, Inferencing, Analogies, Synonyms and Antonyms, Agree/Disagree
	Analytic Reasoning	Memory drill, Decision making, Positioning, Sudoku
	Pictorial Reasoning	Connect the dots, Mazes, Picture Sequence, Spot the difference, etc

Workbook # 1 		
	Verbal Reasoning	Finding the truth, Inferencing, Analogies, Synonyms and Antonyms, Agree/Disagree
	Analytic Reasoning	Sorting, Positioning, Picking, Assorted problems, Numeric and Alphabetic Sudoku
	Pictorial Reasoning	Picture Sequence, Spot the difference, Odd picture

Workbook # 2 		
	Verbal Reasoning	Finding the truth, Classification, Direct and Inverse relationship, Inferencing, Analogies, Agree/Disagree
	Analytic Reasoning	Sequencing, Scheduling, Strategy, Picking, etc
	Pictorial Reasoning	Picture Analogy, Odd picture, Pattern matching, etc

For Grade 3, Grade 4, and Grade 5

Workbook # 3 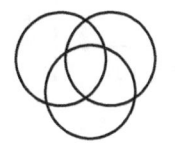		
	Verbal Reasoning	Not, And, Or, If .. then, Conditional inferencing, Unconditional inferencing, Symbolic Logic
	Analytic Reasoning	Lists, Sequencing, Grouping, Venn Diagrams, Graph logic, Number logic, Letter logic, Sudoku
	Pictorial Reasoning	Picture sequence, Picture analogy, Odd picture, Picture difference, Pattern matching

Workbook # 4 		
	Verbal Reasoning	Contradiction, Converse, Inverse, Contrapositive, Conditional inferencing, Symbolic Logic
	Analytic Reasoning	Scheduling, Looping, FIFO, LIFO, Correlation, Venn Diagram, Graph logic, Number logic, Sudoku, etc
	Pictorial Reasoning	Picture sequence, Picture analogy, Odd picture, Picture difference, Pattern matching

Workbook # 5 		
	Verbal Reasoning	Biconditional, Categorical inferencing, Cause and Effect, Symbolic Logic, Agree/Disagree, Word and Sentence analogy
	Analytic Reasoning	Correlation, Grouping, Venn Diagrams, Graph logic, Number logic, Letter logic, Sudoku, etc
	Pictorial Reasoning	Picture sequence, Picture analogy, Odd picture, Picture difference, Pattern matching

********* Essential resource for everyone *********
*http://www.giftoflogic.com *sales@giftoflogic.com

GIFT OF LOGIC™
CRITICAL THINKING & LOGICAL REASONING CURRICULUM
12 WORKBOOKS TO BOOST YOUR THINKING SKILLS

For Grades 6-12, College/University Students, Adults

Primer / **Prereq**

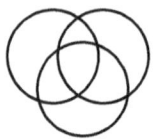

Verbal Reasoning	Logical Operators, Conditional, Categorical and Causal reasoning, Validity, Fallacies, Symbolic Logic
Analytic Reasoning	Positioning, Grouping, Sudoku
Pictorial Reasoning	Pattern perception, Figure formation, Paper folding and cutting, Figure matrix, Rule detection

Workbook# 6

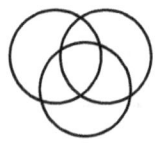

Verbal Reasoning	Arguments-Main point, Must be true, Cannot be true
Analytic Reasoning	Positioning, Grouping, Sudoku
Pictorial Reasoning	Pattern perception, Figure formation, Paper folding and cutting, Figure matrix, Rule detection

Workbook# 7

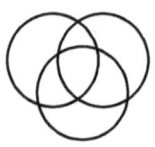

Verbal Reasoning	Arguments-Strengthening, Weakening
Analytic Reasoning	Positioning, Grouping, Sudoku
Pictorial Reasoning	Pattern perception, Figure formation, Paper folding and cutting, Figure matrix, Rule detection

Workbook# 8

Verbal Reasoning	Arguments - Controversy, Paradox
Analytic Reasoning	Positioning, Grouping, Sudoku
Pictorial Reasoning	Pattern perception, Figure formation, Paper folding and cutting, Figure matrix, Rule detection

Workbook# 9

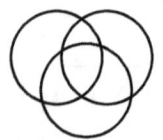

Verbal Reasoning	Arguments- Assumptions, Reasoning strategy
Analytic Reasoning	Positioning, Grouping, Sudoku
Pictorial Reasoning	Pattern perception, Figure formation, Paper folding and cutting, Figure matrix, Rule detection

Workbook# 10

Verbal Reasoning	Arguments-Flawed reasoning, Analogous reasoning
Analytic Reasoning	Positioning, Grouping, Sudoku
Pictorial Reasoning	Pattern perception, Figure formation, Paper folding and cutting, Figure matrix, Rule detection

********* Essential resource for everyone *********
Get the GIFT OF LOGIC™ today !
*http://www.giftoflogic.com *sales@giftoflogic.com

Dear Reader:

Your decision to purchase this book is commendable. You now have in your hands, a comprehensive, easy-to-read book in Critical thinking and Logical reasoning that will introduce you to three different areas of thinking and reasoning - Verbal, Analytical and Pictorial. Solving problems in Verbal Reasoning is important to develop a critical mind. Solving problems in Analytic Reasoning is important to develop a flexible and resourceful mind. Solving problems in Pictorial Reasoning is important to develop a visually alert mind.

This book is presented in a workbook format to help you progress quickly. Parents and teachers are urged to complete the exercises ahead of the student and assist them whenever necessary with the help of detailed answers provided at the end of the book. This book can be used as a supplementary resource in the regular class room or it can be used during winter and summer vacations. College/University students, working professionals and retired individuals will also find the Gift Of Logic(tm) Series very useful in enhancing their problem solving abilities, confidence and general intellect.

Critical thinking and Logical reasoning must be practiced consistently to develop strong cognitive skills. After completing the exercises in this book, continue to read the other books in this series to get familiar with different types of Logical reasoning problems.

This workbook is one in a series of twelve workbooks. Please refer to the brochure before this page for a brief description of each workbook. Visit the website http://www.giftoflogic.com for more information.

                                                                    Happy thinking and reasoning!

TABLE OF CONTENTS

Verbal Reasoning

Flawed Reasoning..8
Analogous Reasoning...23

Analytical Reasoning

Sudoku...46
Positioning...51
Grouping and positioning.. 61

Pictorial Reasoning

Patter perception...68
Figure formation..70
Paper folding and cutting..71
Figure matrix...72
Rule detection...73

Answers

Verbal..77
Analytic...117
Pictorial..140

Certificate of Completion

Name _____ Date _____

VERBAL REASONING

Name _____   Date _____

FLAWED REASONING

In this section on "Flawed Reasoning", you will develop the ability to identify flaws in arguments. Flaws are also called "fallacies".

You will be presented with an argument that has a flaw in its reasoning. Then, you will be posed a question similar to the ones below:
  * The flaw in the argument's reasoning is..
  * The reasoning in the argument is flawed because..

A correct answer is one that accurately describes the flaw in the argument. Incorrect answers do not describe the flaw accurately.

Arguments are flawed for several reasons. These were discussed in the book titled "Critical thinking & Logical reasoning Primer". Understanding the concepts discussed in that book is essential for answering the questions in this section. Some of the common types of flaws are listed below for your quick reference. Consult the "Primer" for a detailed discussion on these and other types of flaws.

   * Flaws in conditional arguments
   * Flaws in categorical arguments
   * Flaws in causal arguments
   * Ad-Hominem arguments
   * Circular arguments
   * Arguments that appeal to authority, appeal to emotion, etc
   * Flaw in composition
   * Flaw in division
   * Flaw in generalization

1   **FLAWED REASONING**   Politician

Andrew: Mr. Pollock is a good politician. He has brought in a lot of jobs to his state. He has also reduced the taxes for people. He deserves a social service award for his contributions.

Brian: Mr. Pollock is a very cunning politician. He is known for his cheating abilities. He has reduced the taxes only because he wants to get elected again this year. It is clear that he should not receive any award for his misdeeds.

Which one of the following most accurately describes a flaw in Brian's argument?

A) it agrees with Andrew's argument that Mr. Pollock deserves an award for his service.
B) it attacks Mr. Pollock's character instead of examining the good things that he has done for his state.

2   FLAWED REASONING   Architect

Amber, Architect: I have won several medals for creating impressive architectural designs. Based on my experience and accomplishments, I would like to recommend that we build an arch at the front entrance of our new city hall building, since arches have a welcoming characteristic to them.

Adrian, Architect: While I agree that arches are majestic at the front entrance, having arches at every entrance to the city hall will make the city hall appear like an ancient structure that is out of touch with modern times. Therefore, your recommendation for the city hall is not feasible to implement.

The reasoning in Adrian's argument is fallacious because the argument

A) distorts Amber's opinion in order to discredit it.
B) appeals to emotion instead of examining the facts.

3 FLAWED REASONING — Science fiction

Only a small number of people like science-fiction movies. These people have a strong interest in science subjects. Therefore, it can be concluded that only a few people like science-fiction movies.

Which one of the following most accurately describes a flaw in the argument's reasoning?

A) it employs a circular argument to prove its claim.
B) the conclusion is clearly stated.
C) the premises provide irrelevant information to reach the conclusion.

4   FLAWED REASONING   Advertisement

Advertisement: Buy the newly introduced RoboVac – a robotic vacuum cleaner for yourself, or gift it to your friends and family. You will be very excited about this purchase. This will definitely be the best thing that you have ever owned. You will be the envy of your friends and family. Their positive opinion of you will increase threefold, thanks to your purchase. You will be the happiest person in this world.

The reasoning in the argument is fallacious because the argument

A) provides premises that are relevant to the conclusion.
B) appeals to emotion instead of providing evidence about RoboVac.
C) employs circular reasoning technique to make its claim.

5   FLAWED REASONING   Psychic

He is the best psychic in the whole world. He has amazing supernatural powers. He can talk to dead people and can foretell your future. He can tell you where you will be living ten years from now, and also what dress you will be wearing next week. Therefore, people must get treated by him for their medical problems.

Which one of the following statements most accurately describes the flaw in the argument's reasoning?

A) it attacks the character of the psychic to justify its reasoning.
B) it appeals to the authority of the psychic to justify its conclusion.

6   FLAWED REASONING   Volunteer

Travis: Mom, it was fantastic that you won the best volunteer award for this month. All my friends congratulated me for your award. I feel very gifted to have you as my mother. Your name will be listed on the school bulletin board forever. You are no doubt the best mom in the entire world. So, we should go to Disney World next month.

Which one of the following statements most accurately describes the flaw in the argument's reasoning?

A) it uses flattery as a tool to achieve the desired result.
B) it uses sympathy as a tool to achieve the desired result.
C) it uses fear as a tool to achieve the desired result.

7   FLAWED REASONING   Traffic Cop

Traffic Cop: You made an illegal left turn at the traffic signal. You can make a left turn only if you see a left-turn arrow sign. By making the left turn, you put several other people at risk. So, I am going to ticket you for this mistake.

Car Driver: I make this left turn every day. If you watch the traffic at this signal, hundreds of cars make this exact left turn without waiting for the left-turn arrow. So, the turn that I made is not illegal.

The car driver's argument is fallacious because

A) it uses flattery to dissuade the traffic cop from issuing a ticket.
B) it attacks the character of the traffic cop to prove its point.
C) it appeals to common practice to justify its conclusion.

8 — FLAWED REASONING — Chocolate

Tom: I don't like chocolate milk. It has a bitter taste to it.

Dick: I love chocolate milk. It is very tasty.

Harry: I also love chocolate milk. It gives me a lot of energy.

Tom: Since all of you like chocolate milk, I like it too.

Which one of the following most accurately describes the flaw in the Tom's line of reasoning?

A) he uses flattery to change his opinion on chocolates.
B) he appeals to the emotions of his friends.
C) he joins the bandwagon to avoid being singled out.

9   FLAWED REASONING   Classroom

Teacher: I am very excited. I will be teaching in a new classroom starting next week.

Student: Why are you excited? Will the new classroom improve the student scores?

Teacher: Sure it will. The new classroom has a new carpet, new tables and chairs. Since everything is new, you will score better than before in your exams.

The fallacy in the teacher's argument is that

A) it appeals to flattery to prove its point.
B) it appeals to novelty to prove its point.

Name _____   Date _____

10   FLAWED REASONING   Criminal

He is a hardened criminal who has been to jail several times. But, he has done a few good things as well. Therefore, he is a good person.

The fallacy in the argument is that

A) it provides irrelevant information to support the conclusion.
B) the conclusion does not take all the premises into consideration.

11   FLAWED REASONING   Party

Andrei: I am throwing a birthday party this weekend. You are welcome. Vladimir is a great friend of mine. So, I am going to invite him as well.

Sergei: Thanks for your invitation. I will surely come. Don't you remember that Vladimir did not invite you to his birthday party last month? So, you must not invite him to your party as well.

Andrei: That's correct, I forgot that entirely. So, I will not invite Vladimir to my birthday party.

The fallacy in the Andrei's argument is that

A) it appeals to authority to support its conclusion.
B) it appeals to spite to support its conclusion.

12   FLAWED REASONING   Team

Nikki: I am going to team up with Cindy and Sammy to do the science project that is due next week. I hope they will actively participate.

Vicki: You know, the last time I teamed up with them to do a project, they did not even show up for the team meetings. Later, when we presented the project to the entire class, they claimed all the credit for the work that I did.

Nikki: What you are saying is correct. They will not do any work, but will take credit for my work. So, I will not team up with them for my science project.

The fallacy in Nikki's argument is that

A) she changes her opinion because of the guilt of associating with Cindy and Sammy.
B) she is naive to trust Cindy and Sammy.

Name _____   Date _____

13 FLAWED REASONING   Celebration

Last year, two hundred students from the city attended the independence day celebrations. Therefore, there will be two hundred students attending the independence day celebrations this year as well.

The fallacy in the above argument is that

A) it assumes that what was true in the past will be true in the future.
B) it provides irrelevant data to support the conclusion.

14 FLAWED REASONING   Eggs

The eggs in this shelf are broken. Therefore, the eggs in all the shelves are broken.

The fallacy in the above argument is that

A) it takes a feature of one member of a group and applies it to the entire group.
B) it takes a feature of an entire group and applies it to one member of the group.

15 FLAWED REASONING — Soccer

Jane: This soccer team has several great players. So, Michael, who is in this team, is also a great player.

The fallacy in the above argument by Jane is that

A) it takes a characteristic of a group and applies it to one member.
B) it provides a conclusion that is in contradiction with its premises.

16 FLAWED REASONING — Postman

A frog jumps randomly from one place to another. A postman goes sequentially from one home to another. Thus, a frog's movement is similar to a postman's movement.

The fallacy in the above argument is that

A) it improperly compares the random movement of a frog with the sequential movement of a postman.
B) it compares two similar movements and concludes that they are similar.

Name _____ Date _____

17 FLAWED REASONING   Profession

Jason graduated high school with good academic credentials. So, he now must choose between the Engineering and Medical professions.

The fallacy in the above argument is that

A) it assumes that he does not have any other choice except the Engineering and Medical professions.
B) the premise is totally irrelevant to the conclusion.

18 FLAWED REASONING   Demonstration

Nearly one hundred people gathered at the Innovation stadium to see a demonstration of a robotic-chef by Mr. Geekman. He claimed that his robot has the capability to cut vegetables. But, during the demonstration, the robot's hand broke, and the demonstration was cancelled. Therefore, Mr. Geekman has made a false claim about the capabilities of the robot.

The fallacy in the above argument is that

A) it fails to consider the fact that Mr. Geekman has a good reputation in the community.
B) is assumes that if a claim is not proved once, it is a false claim.

19   FLAWED REASONING   Election

The Greenleaf party is certain that they will win this year's election by a large majority. They cite the results of a survey of their party's supporters. Ninety percent of their supporters said that they will win and therefore the party believes that they will get ninety percent of all the votes in the election.

The fallacy in the above argument is that

A) it takes the results from an unbiased sample and applies it to the entire voting population.
B) it takes the results from a biased sample and applies it to the entire voting population.

Name ———————————— Date ————————————

20 FLAWED REASONING — Competition

The principal of a high school was tasked with selecting a team of six students without bias to represent the school in the state spelling competition. The school has 200 girls and 400 boys. In the two spelling competitions during prior years when the school won medals, there were two girls and four boys in the team. So, in order to select a random sample of six students, she picked two girls and four boys to represent the school this year as well.

The fallacy in the principal's method of selecting the team is that

A) it does not provide a rationale to justify its decision.
B) it predetermined the composition of the sample.

21 FLAWED REASONING — Restaurant

Inspector: Last month, I ordered you to upgrade all the food processors in your kitchen. So, you need to show me proof that you have followed my order.

Restaurant Owner: Here are the receipts showing that I made all the upgrades that you requested.

Inspector: I am not satisfied. So, I will impose a heavy fine on you.

The fallacy in the Inspector's decision to impose a fine is that
A) it attacks the owner's character.
B) it willfully ignores the evidence presented.

22   FLAWED REASONING   Hybrid

Sabrina: The new Zoro hybrid car in the market has a very bad emission problem. It emits more nitrogen oxide than is permitted by the environmental agency. Therefore, it is very harmful to be near one of these cars.

Mariana: Myself and my father love this car very much. We test drove this car for five miles and it runs very smoothly. It can hold seven musical discs and has a Global Positioning System. The best thing about this car is that it runs forty miles per gallon. It has some emission problem, but it is thrilling to drive this car.

Sabrina: I love the features that you describe, Mariana. So, I will buy one Zoro for myself this week.

The fallacy in Sabrina's argument is that

A) it appeals to popular appeal to reach a conclusion.
B) it ignores unfavorable evidence and considers only the favorable evidence.

Name _____   Date _____

23   FLAWED REASONING   Detectives

Several teams of police detectives were hot in trail of a criminal. This person has robbed several banks in the last five years. He also taunts the detectives by sending letters and making anonymous calls to news media about his accomplishments. So, when it was reported on television that a man with a long criminal history has been arrested, the detectives heaved a sigh of relief and declared the case solved.

The fallacy in the above argument is that

A) it jumps to a conclusion with little or no evidence.
B) it bases its decisions on irrelevant pieces of information.

24   FLAWED REASONING   Commuters

To reduce overcrowding in trains, nearly ninety percent of commuters think that more trains must be introduced into the city's mass transit system. Therefore, the mayor decided to introduce more trains.

The fallacy in the above argument is that

A) it appeals to emotion rather than providing facts to support its conclusion.
B) it appeals to majority's opinion to make its conclusion.

Name _____   Date _____

| 25 | FLAWED REASONING   Article |

Newspaper Article: We recommend Mr. Harry to the state legislature and request everyone to vote for him in this month's elections. He is the son of a poor farmer and has worked hard all his life. He does not have a lot of money to advertise his position in television. He donates most of his income to charities and lives a very simple life.

Which one of the following most accurately describes a flaw in the argument's reasoning?

A) it appeals to the pity of the readers to win support for its conclusion.
B) it resorts to cherry-picking to prove its conclusion.

| 26 | FLAWED REASONING   Car |

The Y-series of the Zoro line of hybrid cars are excellent fuel savers. It is very expensive due to the very advanced technology that is used to inject fuel into the engine. These cars are expected to be used by very affluent people. Therefore, we expect these cars to sell in large numbers because of their affordability.

The fallacy in the above argument is that

A) the premises in the argument contradict the conclusion.
B) it appeals to envy to reach its conclusion.

27   FLAWED REASONING   Award

Pam: I want to give the Biggest Eater award to someone in our school. Do you think you qualify?

Sam: Easily! I can eat a hundred pancakes and fifty sodas in one sitting. No one else can eat this much. Give me the award!

Pam: Ok. I hereby confer upon you the Biggest Eater award for this year.

The fallacy in Pam's argument is that

A) it does not provide a rationale to justify its decision.
B) it relies on hyperbole to infer its conclusion.

Name —————————————— Date——————————

ANALOGOUS REASONING

In this section on "Analogous Reasoning" you will develop the ability to identify the analogy between two arguments. Analogous arguments are also called Parallel arguments.

You will be given an argument that has a particular method of reasoning. Your task will be to find one argument from a list of arguments that is analogous (parallel) to the given argument.

After the argument is presented, questions are posed as follows:
* The pattern of reasoning in the argument above is similar to which one of the following arguments?
* Which one of the following arguments is parallel in its structure to the above argument?

The correct answer is the one that has the same type of premises and the same type of conclusion as the given argument. The two arguments must be analogous in structure, but they do not have to use the exact same words. If the given argument has a flaw in reasoning, the analogous argument also must have a flaw in reasoning. If the given argument uses conditional reasoning, the correct answer must also have conditional reasoning. Incorrect answers are those that are not analogous in premises or conclusion or both.

Example: (analogous premises and analogous conclusion)
Argument 1: If he is wise, he must not jump from the bridge. He is wise and so, he did not jump from the bridge.
Argument 2: If she is nice, she must not lock the door. She is nice and so, she did not lock the door.

Verbal Reasoning
© Gift Of Logic, Inc * Copying prohibited

1   ANALOGOUS REASONING   Iron

All the cars, when considered together, use a lot of iron. So, Stanley's car also uses a lot of iron.

The flawed pattern of reasoning in the argument above is similar to that in which one of the following?

A) Joan loves democracy. So, the people of her country also love democracy.
B) All the fruits are healthy to eat. So, the apples are healthy to eat.
C) All the trucks, when considered together, pollute the air very much. So, Nancy's truck also pollutes the air very much.

| 2 | ANALOGOUS REASONING | Talent |

If a person plays the piano, that person must be talented. Tina plays the piano and therefore she must be talented.

The reasoning in the argument above is parallel in its structure to which one of the following?

A) If a person drives a car, that person must be a good driver. Sam does not drive a car and so he must not be a good driver.
B) If a person draws well, that person must be creative. Todd draws well and consequently he must be creative.

3   ANALOGOUS REASONING   Studies

Ian's mom told him that if he studies for two hours today, then he can eat an ice-cream. Ian ate an ice-cream today; so he must have studied for two hours today.

The flawed pattern of reasoning in the argument above is similar to that in which one of the following?

A) If the train maintains its speed, it will arrive at the station in time. The train arrived in time; so it must have maintained its speed.

B) If it snows, then the temperature will be cold. So, since it did not snow, the temperature was not cold.

4   ANALOGOUS REASONING   Fire ants

Before going to the backyard, Vivian was warned by his dad that he should be careful about fire ants in the yard. Vivian claimed that since the home was new, the backyard will not have fire ants. So, he ignored the warning.

The reasoning in the argument above is analogous to which one of the following?

A) Ronald was alerted by his mom not to get wet in the rain since he could damage his watch. But, Ronald reasoned that his watch was water proof. Therefore, he dismissed the warning.

B) Amber's mom cautioned her to be careful while handling the electric drill. Amber argued that the new drill was safe. So, she ignored the warning to be cautious.

5 ANALOGOUS REASONING — Syllogism

All P are Q.
All Q are R.
Therefore, all P are R.

Which one of the following arguments has a reasoning similar to the argument above?

A)
All M are N.
If there is a N, then it is a O.
Therefore, all O are M.

B)
All A are B.
If there is a B, then it is a C.
Therefore, all A are C.

6 — ANALOGOUS REASONING — Blood pressure

A doctor attending to a patient first noticed that he had a high temperature, and after some time, a high blood pressure. So, the doctor concluded that the patient's high temperature caused the high blood pressure.

The pattern of reasoning in the argument above is similar to that in which one of the following?

A) The home owner noticed that when he walked up the stairs, the stairs swayed first, and then there was a squeaking noise. So, he concluded that the squeaking noise made the stairs to sway.

B) When Lucy was driving her car, she first noticed a blue van and then after a minute, she witnessed an accident. So, she concluded that the blue van caused the accident.

Name ———————————————   Date ———————————————

7   ANALOGOUS REASONING   Cause and Effect

P, Q, and R are events that happened at different times. P caused Q. The effect of Q is R. Therefore, R was caused by P.

Which one of the following arguments has a reasoning similar to the argument above?

A) L, M, and N are events that happened at different times. L caused M to happen. M triggered N. Therefore, L is the effect of N.
B) I, J, and K are events that happened at different times. J is the effect of I and J is the cause of K. Therefore, K is the effect of I.

8 ANALOGOUS REASONING — Theft

Of late, incidences of theft of watches have increased. So, people must prove ownership of their watches when requested.

Which one of the following arguments has a reasoning similar to the argument above?

A) In recent days, cars are being stolen in record numbers. Consequently, anyone must be able to prove ownership of their vehicles when asked to do so.

B) Lots of computers are being stolen these days. Therefore, people must show proof of ownership of their computers on request.

| 9 | ANALOGOUS REASONING | Blame |

Rhonda was sure that facial cream will help remove her pimples and applied a lot of it on her face. But, the pimples did not go away. So she blamed her pimples for the problem.

The pattern of reasoning in the argument above is similar to which one of the following?

A) Alex was sure that the ant spray that he used will kill the ants in the kitchen. But, even after spraying a lot, the ants did not die. So, Alex blamed the ants for the problem.

B) Bruce kept wiping the table hard with a grease remover because he wanted to get rid of the sticky substance on the table. The sticky substance did not go away. So, Bruce blamed the grease remover for the problem.

10 ANALOGOUS REASONING — Radio

Most people like to lead peaceful lives. Hence, radio programs that deal with controversies must not be aired.

The reasoning in the argument above is analogous to which one of the following?

A) Most children like to eat healthy food. So, food items that are spicy must not be served to them.

B) Most people like humor. So, television programs that show serious content must not be telecast.

Name _____   Date _____

11 ANALOGOUS REASONING — Bank loan

It is painful to get a bank loan. They ask for a lot of information. So, the loan application process must be revamped.

Which one of the following arguments has a reasoning similar to the argument above?

A) It is a hassle to clear the security at the airport. They check a lot of things. So, the security process must be modified.

B) It is difficult to rent a car. You never get the car that you want. So, the car renting process must be revised.

| 12 | ANALOGOUS REASONING | Emergency |

The pilot of an airplane decided to make an emergency landing and requested the control tower for help. Hence, the traffic controllers decided that the plane was having technical difficulties.

The flawed pattern of reasoning in the argument above is similar to that in which one of the following?

A) A man in an air balloon requested help to land. So, the helpers on the ground decided that he was not feeling well.
B) The helicopter pilot requested help to land immediately. Therefore, the traffic controllers decided that there was some problem with the helicopter.

13 ANALOGOUS REASONING — Dresses

Three dress sizes, small, medium and large are available to fit people of all sizes. But, not all people can fit themselves into one of these three sizes. Therefore, some people must hire a tailor to make a dress that fits.

The pattern of reasoning in the argument above is similar to that in which one of the following?

A) Gloves are available in two sizes: small and medium. These sizes may not fit everyone. So, some must hire a glove maker to make one that fits.
B) One has three choices of jeans – straight, pleated and relaxed. One may not find these sizes suitable. So, they must wear the one that fits the best.

14   ANALOGOUS REASONING   Cars

My first car was made by Honda. My second car was also made by Honda. My first car lasted for twenty years. Therefore, my second car also will last for twenty years.

The pattern of reasoning in the argument above is similar to that in which one of the following?

A) My first watch was made by Timex. My second watch was also made by Timex. My first watch lasted for fifteen years. Therefore, my second watch will also last for fifteen years.

B) My first pet was a dog. My second pet was also a dog. My first pet lived for ten years. Therefore, my second pet will live for twenty years.

15 ANALOGOUS REASONING — Cooking

Jack loves cooking. Jack and Jill have the same parents. Therefore Jill also loves cooking.

The pattern of reasoning in the argument above is similar to that in which one of the following?

A) Martha likes music. Martha and Maggie have a sister in common. So, Maggie also likes music.

B) Rick likes Science. Rick and Sam have a brother in common. So, Sam likes History.

ANALYTICAL REASONING

1 SUDOKU

Solve the following Sudoku. A correctly solved Sudoku has numbers 1-9 appearing only once in each row, each column and each 3x3 grid. Solving Sudokus will help you to gain valuable analytic skills.

	6		7	2	5		1	
	8			1			6	
1	2	7		8		5	3	4
6	9	1	2	4	8	3	7	5
	4			6			9	
2	3	8	5	7	9	1	4	6
9	7	6	1	5	2	4	8	3
8		2	6		4	9		7
3		4	8		7	6		1

Name _____ Date _____

2 SUDOKU

Solve the following Sudoku. A correctly solved Sudoku has numbers 1-9 appearing only once in each row, each column and each 3x3 grid. Solving Sudokus will help you to gain valuable analytic skills.

	3			9			8	
9	8	7	4	1	6	3	5	2
5		2	3		7	9		1
8	7	3	5	2	4	1	9	6
6		9	1		8	5		5
1	4		9	6		2	7	
	9	4		5	1		6	3
3		8		4	2		1	9
7	1		8	3		5		4

Analytical Reasoning   Answers-118
© Gift Of Logic, Inc * Copying prohibited

Name _____ Date _____

3 SUDOKU

Solve the following Sudoku. A correctly solved Sudoku has numbers 1-9 appearing only once in each row, each column and each 3x3 grid. Solving Sudokus will help you to gain valuable analytic skills.

	6		8		2	5		3
4		3	6		9	1		7
2	1		4		7	8		6
	9	8	1		5		6	
5		4	2	8			7	1
1	2		6		9	3	4	5
8		2	9	6		7		5
6	5	1	3		4		8	9
3	7			2	8	6	1	4

Analytical Reasoning

Name _____ Date _____

4   SUDOKU

Solve the following Sudoku. A correctly solved Sudoku has numbers 1-9 appearing only once in each row, each column and each 3x3 grid. Solving Sudokus will help you to gain valuable analytic skills.

5	6	2	7	8	1	9	3	4
1		7	9		6	2		8
	4	8		2	3		6	1
7	1	4	8	6	5	3	9	2
8		5	1		9	4		6
	9	6		7	2		1	5
2	7	3	6	5	8	1	4	9
6		9	3		4	5		7
	5	1		9	7		8	3

Analytical Reasoning
© Gift Of Logic, Inc * Copying prohibited

Name _____   Date _____

5   SUDOKU

Solve the following Sudoku. A correctly solved Sudoku has numbers 1-9 appearing only once in each row, each column and each 3x3 grid. Solving Sudokus will help you to gain valuable analytic skills.

	8	4	2	5	9	6	1	3
6	3	5	7		8		2	4
9	1	2	6	4	3	5	7	8
5	6	9	8	2	4		3	7
1		8	3		6	2	9	5
2	7	3	5	9	1	8	4	6
3		1	4	6	2	7		9
4		7				3		1
8		6	1	3	7	4		2

Analytical Reasoning   Answers-121
© Gift Of Logic, Inc * Copying prohibited

| 1 | POSITIONING | vacancy |

SCENARIO

Three boys Andrew, Bashir, and Chen must be seated in a row of four chairs. Andrew must sit immediately to the left of Bashir.

QUESTIONS

1) If the fourth chair must be left vacant, list all the possible ways in which the three boys can be seated?

2) If the third chair must be left vacant, then which of the following must be true?

    A)   Chen must sit in the first or second chairs.
    B)   Chen must sit in the first or third chairs.
    C)   Chen must sit in the fourth chair.

3) If the second chair is not occupied, then which of the following is possible?
    A)   Andrew, Vacant, Bashir, Chen
    B)   Chen, Vacant, Andrew, Bashir
    C)   Bashir, Chen, Vacant, Andrew

Analytical Reasoning   Answers-122
© Gift Of Logic, Inc * Copying prohibited

Name _____   Date _____

2   POSITIONING   vacancy, someone

SCENARIO

Three girls Anita, Beena, and Cathy must be seated in a row of four chairs, numbered 1,2,3 and 4. Someone must sit in chairs 2 and 3. Anita must sit immediately to the left of Beena.

QUESTIONS

1) Which of the following seatings are not possible?

    A) Anita, Beena, Vacant, Cathy
    B) Anita, Beena, Cathy, Vacant
    C) Vacant, Anita, Beena, Cathy
    D) Cathy, Vacant, Anita, Beena

3 SCENARIO

Three girls Anita, Beena, and Cathy must be seated in a row of four chairs, numbered 1,2,3 and 4. Someone must sit in chairs 2 or 3. Anita must sit immediately to the left of Beena.

QUESTIONS

1) Which of the following seatings are possible?

    A) Anita, Beena, Vacant, Cathy
    B) Vacant, Cathy, Anita, Beena
    C) Vacant, Beena, Anita, Cathy
    D) Cathy, Vacant, Anita, Beena

Analytical Reasoning   Answers-123
© Gift Of Logic, Inc * Copying prohibited

Name —————————————— Date ——————————

| 4 | POSITIONING | identify vacancy |

SCENARIO

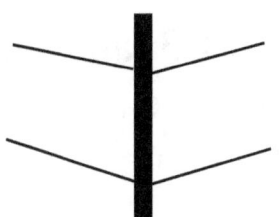

There are four branches in a tree, two on the left side and two on the right side, as shown. Three birds, A, B, and C fly into this tree to sit on the branches. Only one bird can sit in each branch.

QUESTIONS

1) If B and C must sit in the right side of the tree, and C must sit below B, then A must sit opposite to B.

$$\text{A) True} \quad \text{B) False}$$

2) If C and B can sit on either side of the tree, and C must sit below B then which one of the following must be true?

   A) Any branch can be vacant.
   B) Bird A must always sit on a top branch.

5 POSITIONING — rank

SCENARIO

There are four types of cheese in the kitchen, C1, C2, C3, and C4. Three rats, R1, R2, and R3 are ready to eat the cheese. Only one type of cheese can be eaten by any one rat. R1 can eat C1 or C3, R2 can eat C2 or C4. R3 can eat C3 or C1.

QUESTIONS

1) If R1 eats C1 and R2 eats C2 then which of the following must be true?

   A) R3 eats C1.
   B) R2 eats C4.
   C) R3 eats C3.

2) If R1 eats C1 and R2 eats C4, then which of the following must be true?

   A) R3 eats C3 and C2 is not eaten.
   B) R3 eats C1 and C2 is not eaten.

Name _____   Date _____

6 POSITIONING   few in many spots

SCENARIO

A doctor has four appointments – two in the morning, and two in the afternoon. Three patients, A, B, and C want to see the doctor. At least one morning and at least one evening appointment must be filled.

QUESTIONS

1) The number of ways appointments can be scheduled so that patient A is seen first in the morning and patient B is also seen in the morning is

    A) 1   B) 2   C) 3   D) 4

2) The number of ways appointments can be scheduled so that only patient B is seen in the afternoon is

    A) 2   B) 3   C) 4

3) The number of scheduling possibilities in which only patient A can meet the doctor in the morning and patient B is the first to see the doctor in the afternoon is

    A) 2   B) 3   C) 4

7 POSITIONING — many in few spots

SCENARIO

Four sailors A, B, C, and D are to be assigned to three boats. One sailor only in each boat. If C is assigned to sail, D must sail as well.

QUESTIONS

1) If A sails, then which sailor will not be able to sail?
   A) A or B   B) B or C   C) C or D

2) If D does not sail, then which of the following will not sail?
   A) A   B) B   C) C

3) If D does not sail, then the number of unassigned boats will be
   A) 2   B) 1   C) 3

POSITIONING — many in few spots

SCENARIO

Four patients, A, B, C, and D need appointments to see a doctor this week. The doctor has two appointments available on Monday and one on Tuesday.

QUESTIONS

1) If one appointment on Monday is filled by A and the appointment on Tuesday is filled by C, then which of the following must be true?

   A) If D sees the doctor on the same day as A, then D sees the doctor on Tuesday.
   B) Exactly one among B or D will not able to see the doctor this week.

2) If A and C must be seen on the same day, then which of the following must be true?
   A) B can be seen on the same day as A.
   B) D cannot be seen on the same day as C.

9   POSITIONING   schedule, elimination

SCENARIO

The availability of GreatLife recreation center from Monday to Friday is shown below.

	Monday	Tuesday	Wednesday	Thursday	Friday
morning	open	open	open	open	open
afternoon	open	closed	open	open	closed

Tom can play only one game every day. If he plays in the morning, he must play in the afternoon the next day.

QUESTIONS

1) Can he play on Monday morning?
    A) Yes   B) No

2) Can he play on Friday morning?
    A) Yes   B) No

3) How many mornings can he play in a week?
    A) 5   B) 3   C) 2

10 POSITIONING   vacant spots, logical chaining

SCENARIO

Four people A, B, C, and D must be seated in six spots that are spread in two rows as shown.

The person named A must sit next to B on the same row. The person named B must sit above C. The person named D must sit immediately after C on the same row.

```
              1     2     3
top row      ___   ___   ___

bottom row   ___   ___   ___
```

QUESTIONS

1) If B sits in the top row-second position, then there is one vacant spot in each row.
   A) True   B) False

2) If C sits in the bottom row-first position, then the vacant positions are one below the other.
   A) True   B) False

3) If B sits in the top row-second position and the spot immediately to the left of B is vacant, then D sits in the second row-second position.
   A) True   B) False

Name _____   Date _____

11　　　　　　　POSITIONING　　　　　　　order

SCENARIO

The postman to Dove Creek Lane has to deliver seven packages to three homes A, B, and C. Each home gets at least one package. He must deliver the packages first to the home with the highest number of packages and then to the next in descending order.

QUESTIONS

1) If he delivers one package to A and four packages to C, then the order of his trip to the three homes is

    A) C, B, A　　　　B) C, A, B　　　　C) B, A, C

2) If he delivers to B first and could deliver next to either A or C, then the number of packages he delivers to A, B, C is

    A) A-1, B-4, C-2　　B) A-1, B-5, C-1

Name _____   Date_____

1 GROUPING AND POSITIONING

SCENARIO

Three flowers are to be selected for planting from a group of four flowers of colors red, blue, green, and yellow. The selected flowers are to be planted consecutively in a row.

If the blue flower is selected, then it must be planted after the red flower.

QUESTIONS

1) Which of the following choices satisfy the selection and positioning constraints for planting in the garden?

   A) Blue, Red, Yellow
   B) Red, Blue, Yellow
   C) Red, Green, Blue
   D) Green, Yellow, Red

Analytical Reasoning
© Gift Of Logic, Inc * Copying prohibited

Name ——————————————   Date ——————————————

2 GROUPING AND POSITIONING

SCENARIO

Tracy plans to invite only three families for dinner out of the four that she knows. The families can be invited on Monday, Tuesday, and Wednesday; one family only on any given day. The four families are F1, F2, F3, and F4. Family F2 can be invited on Wednesday only.

QUESTIONS

1) Which of the following invitations can be made for Monday, Tuesday, and Wednesday?

    A) F1,F3,F2   B) F1,F2,F4   C) F1,F3,F4

2) If F2 must be invited, and F1 is also invited, then one of which of the following families can also be invited?
    A) F3 or F4   B) F3 or F1   C) F2 or F3

Name _____   Date _____

3   GROUPING AND POSITIONING

SCENARIO

Sheela is moving into an apartment and is setting up her rooms. She has a TV, a Sofa, a Computer, and a Printer. The Computer and the Printer must be in the same room. If the TV is placed in a room, the Sofa also must be placed in the same room. If the Sofa is placed in a room, the Computer also must be placed in the same room.

QUESTIONS

1) Which of the following assignments of items is correct?

   A) TV, Sofa, Computer
   B) TV, Computer, Printer
   C) Sofa, Computer, Printer

4 GROUPING AND POSITIONING

SCENARIO

Three statues must be picked out from the available four statues that are Red, Green, Blue, and Yellow in color. The selected statues are to be placed in three positions, numbered 1, 2 and 3.

The red and blue statues must be selected.
If the blue statue is placed in the third position, the green statue must be placed in the first position.

QUESTIONS

1) Which of the following combinations of statues can be selected?
   A) Red, Green, Blue
   B) Green, Blue, Yellow
   C) Blue, Yellow, Red
   D) Yellow, Red, Green

2) Which of the following arrangement of statues are valid in positions 1, 2 and 3 respectively?

1	2	3	Valid?
Yellow	Red	Blue	
Red	Green	Blue	
Blue	Green	Red	
Yellow	Blue	Red	
Green	Red	Yellow	

Analytical Reasoning

5 GROUPING AND POSITIONING

SCENARIO

Five monkeys m1, m2, m3, m4, and m5 are to be put in three cages c1, c2 and c3 that are next to each other. Only one monkey can be put in one cage. Monkeys m1, m2, and m3 are tall, and monkeys m4 and m5 are short.

Cage c1 can hold only a tall monkey.
Cage c3 can hold only a short monkey.
Tall monkeys must be before the short monkeys.

QUESTIONS

1) Which of the following are valid assignments of monkeys to the three cages?

Cage c1	Cage c2	Cage c3	Valid?
m1	m3	m4	
m4	m1	m5	
m1	m5	m3	
m2	m1	m4	

2) If cage c2 can hold only a short monkey, then which of the following can be valid assignments of monkeys to the cages?

Cage c1	Cage c2	Cage c3	Valid?
m2	m3	m4	
m2	m4	m5	
m3	m5	m4	
m2	m1	m4	

6 GROUPING AND POSITIONING

SCENARIO

Four birds are to be selected from two groups of birds - Group 1 comprising of birds b1, b2, and b3 and Group 2 comprising of birds b4, b5, and b6. The selected birds are to be placed in four cages labeled c1, c2, c3 and c4 from left to right.

If one cage has a bird from one group, the next cage should have a bird from an alternate group.
If b1 and b3 are selected, then b1 must be in a cage that is before b3's.
If b2 and b6 are selected then b2 must be in a cage that is before b6's.

QUESTIONS

1) Which of the following cage assignments of birds are valid?

Cage c1	Cage c2	Cage c3	Cage c4	Valid?
b2	b1	b4	b5	
b3	b5	b1	b6	
b6	b3	b4	b1	
b1	b6	b2	b4	
b6	b2	b5	b3	
b1	b4	b2	b6	

2) If birds b1 and b4 are not selected, then which of the following cages cannot hold bird b2?
   A) 1   B) 2   C) 3   D) 4

Name ——————————— Date ———————

PICTORIAL REASONING

Name _____   Date _____

PATTERN PERCEPTION - MISSING PATTERN

Find the correct figure from the three alternatives given that will fit logically into the missing portion of the figure on the left.

1   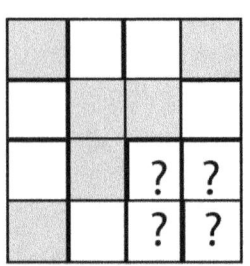 A 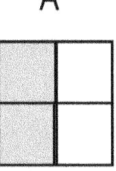   B 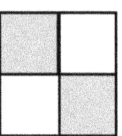   C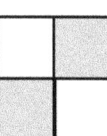

2   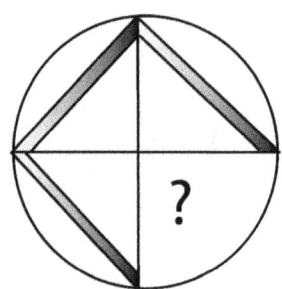 A    B    C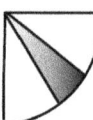

3   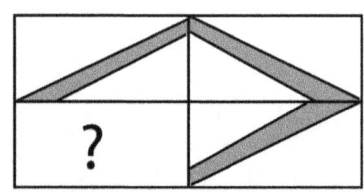 A 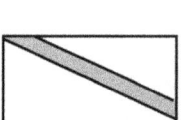   B 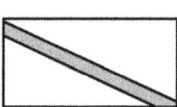   C

4   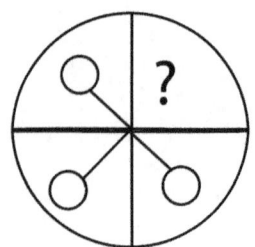 A 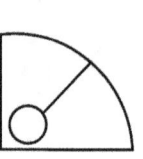   B    C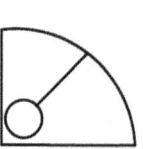

Pictorial Reasoning

Name _____ Date _____

PATTERN PERCEPTION - CONTINUING PATTERN

Find the correct figure from the two alternatives given that will logically continue the pattern of figures on the left.

5

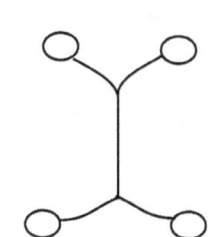 ?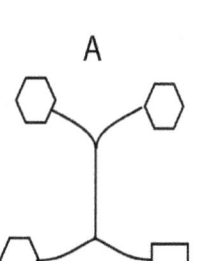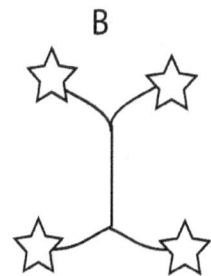

6

 ? (A) (B)

7

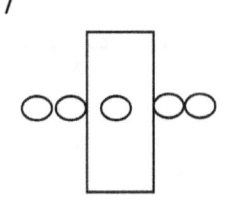 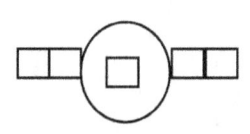 ?

8

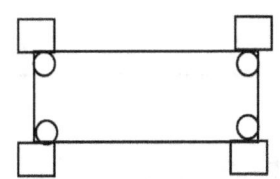 ?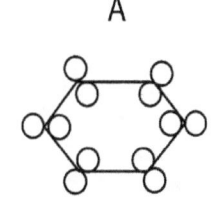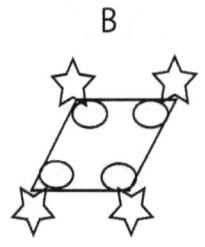

Pictorial Reasoning Answers-140 69
© Gift Of Logic, Inc * Copying prohibited

Name _____   Date _____

FIGURE FORMATION

Find the correct figure that will be formed when the mask shown is applied to the figure on the left.

1

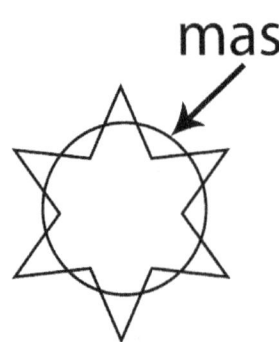 =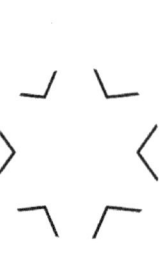

A   B

2

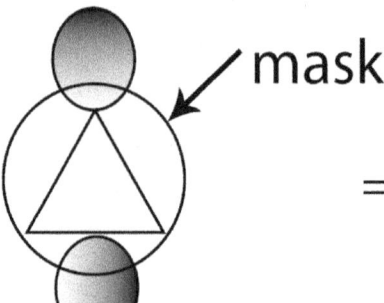 =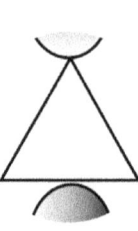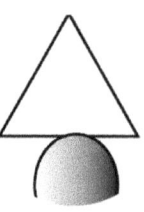

A   B

3

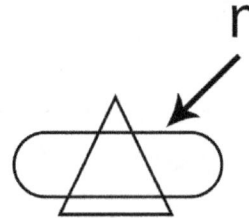 =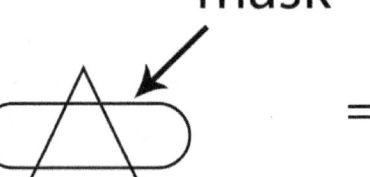

A   B

4

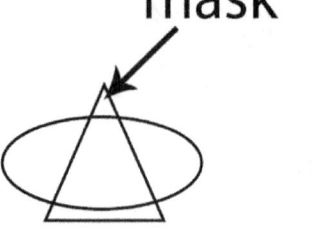 =

A   B

Pictorial Reasoning   Answers-140
© Gift Of Logic, Inc * Copying prohibited

PAPER FOLDING AND CUTTING

Find the correct figure that will be formed when the paper on the left is folded in the direction of the arrows, and then holes are cut in it as shown.

1 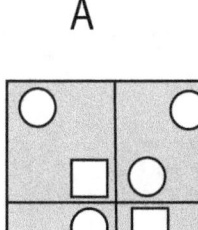   A 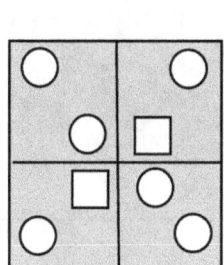   B 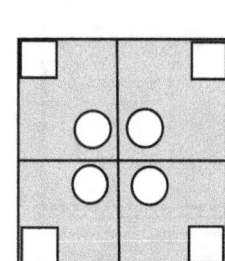   C   D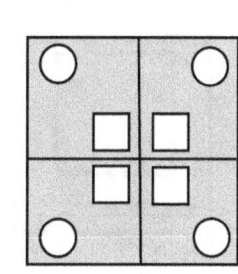

2 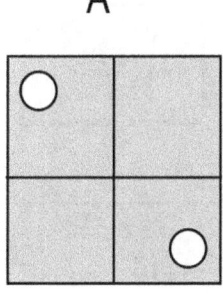   A 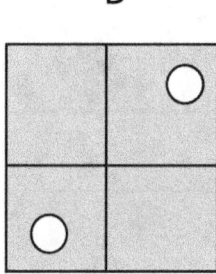   B   C 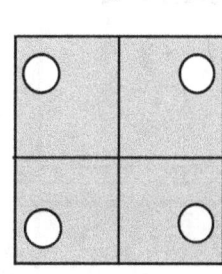   D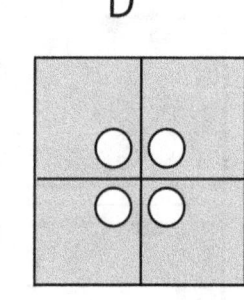

3 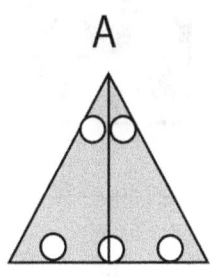   A   B 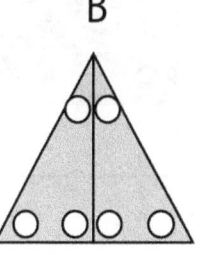   C 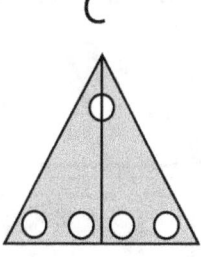   D

Pictorial Reasoning
© Gift Of Logic, Inc * Copying prohibited

Name ————————————————— Date ——————————————

FIGURE MATRIX - ANALOGY

Find the correct figure from the alternatives given that will fit in the empty box such that, the bottom two figures are related in the same way as the top two figures.

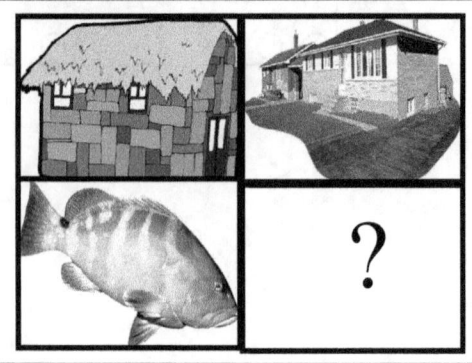
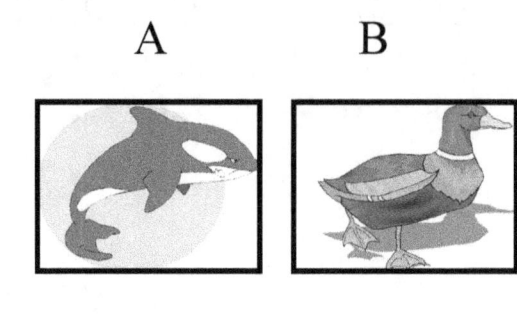

Pictorial Reasoning   Answers-141
© Gift Of Logic, Inc * Copying prohibited

Name —————————————— Date——————————

FIGURE MATRIX - SIMILARITY

Three figures in the 2 x 2 matrix have similar characteristics. Find the fourth figure from the alternatives given that is also alike.

5      A   B   C

6      A   B   C

7      A   B   C

8      A   B   C
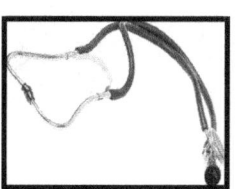  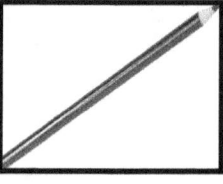

Pictorial Reasoning   Answers-141
© Gift Of Logic, Inc * Copying prohibited

RULE DETECTION

Read the given rule in each question. Then, find the correct choice from the alternatives given that satisfies the rule.

1. Figures move closer to each other

A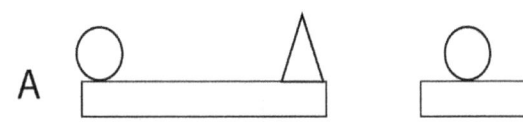

B

2. Each figure is sheared

A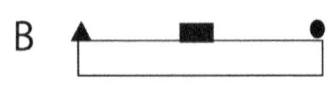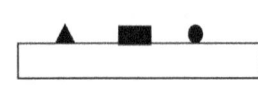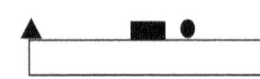

B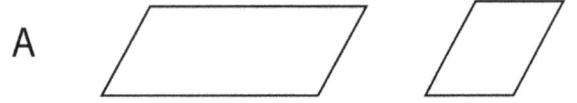

3. Each figure is fully enclosed by another figure

A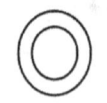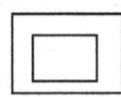

B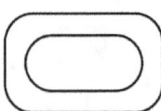

Name ——————————————   Date ——————————

RULE DETECTION

Read the given rule in each question. Then, find the correct choice from the alternatives given that satisfies the rule.

4   Each figure is made of two different shapes

A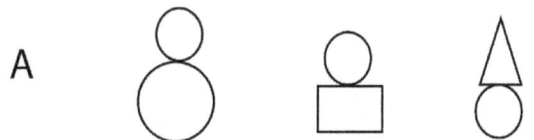

B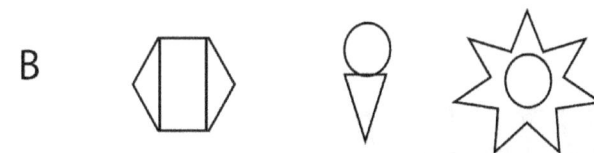

5   Figures are halved to get the next figure

A

B

6   Figures are joined at the edges

A

B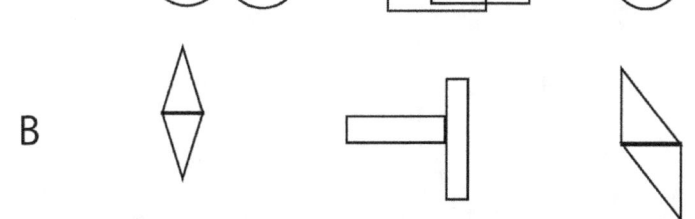

Pictorial Reasoning   Answers-141
© Gift Of Logic, Inc * Copying prohibited

ANSWERS

1 FLAWED REASONING

Andrew: Mr. Pollock is a good politician.
Brian: Mr. Pollock is a very cunning politician.

Which one of the following most accurately describes a flaw in Brian's argument?
A) It agrees with Andrew's argument that Mr. Pollock deserves an award for his service.
B) It attacks Mr. Pollock's character instead of examining the good things that he has done for his state.

ANSWER

Answer: B

A – incorrect – this choice does not describe the flaw in the argument. It also does not agree with Andrew's argument.

B – correct – Brian should assess Mr.Pollock's achievements, but instead attacks his character. This type of reasoning is flawed. The argument contains the "ad-hominem" fallacy.

2	FLAWED REASONING

Amber, Architect: I have won several medals..

The reasoning in Adrian's argument is fallacious because the argument
A) distorts Amber's opinion in order to discredit it.
B) appeals to emotion instead of examining the facts.

ANSWER

Answer: A

A – correct - Adrian's argument distorts the facts to assert its conclusion.

Note that Amber recommends arches only at the front entrance, but Adrian misinterprets it by saying that Amber wants to build an arch at every entrance to the building and discredits it by saying that it is out of touch with modern times. There is a deliberate exaggeration/ distortion in Adrian's argument. This type of fallacy is called the Straw Man fallacy.

B – incorrect- Adrian is not appealing to anyone's emotion.

3 FLAWED REASONING

Only a small number of people like science-fiction..

Which one of the following most accurately describes a flaw in the argument's reasoning?
A) it employs a circular argument to prove its claim..
B) the conclusion is clearly stated.
C) The premises provide irrelevant information to reach a conclusion.

ANSWER

Answer: A

A – correct – the premise of the argument is that only a small number of people like science fiction movies. The conclusion of the argument is also that only a few people like science fiction movies. The premise and conclusion say the same thing in a circular fashion. This flaw is called the flaw of circular reasoning or begging-the-question fallacy. The premise is supposed to lead us to the conclusion, but it claims that the conclusion is true. The conclusion is supposed to be derived from the premise, but instead, it points to the premise again.

B – incorrect – If the conclusion is clearly stated, then why would it be a flaw? The flaw in this argument is that the conclusion is the same as the premise.

C – incorrect. The premises do provide relevant information. So, this choice does not describe the flaw in the argument.

4 FLAWED REASONING

Advertisement: Buy the newly introduced RoboVac..

The reasoning in the argument is fallacious because the argument
A) provides premises that are relevant to the conclusion.
B) appeals to emotion instead of providing evidence about RoboVac.
C) employs circular reasoning technique to make its claim

ANSWER

Answer: B

The argument's conclusion is "Buy the newly introduced RoboVac". Several premises are given, but they all appeal to emotion rather than explaining how this RoboVac will be useful to anyone.

A – incorrect. The premises are not relevant to the conclusion.

B – correct – the argument appeals to emotion, but does not provide sound evidence to make people buy the RoboVac. This type of flaw is called the "appeal to emotion" fallacy.

C – incorrect – the premises do not have a circular relation with the conclusion.

5 FLAWED REASONING

He is the best psychic in the whole world..

Which one of the following statements most accurately describes a flaw in the argument's reasoning?

A) it attacks the character of the psychic to justify its reasoning.
B) it appeals to the authority of the psychic to justify its conclusion.

ANSWER

Answer: B

A – incorrect – the argument does not attack the character of the psychic at all.

B – correct – the argument provides facts about the psychic's powers and concludes that people must get treated by him for their medical problems.

This reasoning is flawed because a psychic is not a medical professional, and so we should not rely on him for medical problems. The argument appeals to his authority to prove its point. This flaw is called the "appeal to authority" flaw.

6	FLAWED REASONING

Travis: Mom, it was fantastic that you ..

Which one of the following statements most accurately describes the flaw in the argument's reasoning?
A) it uses flattery as a tool to achieve the desired result.
B) it uses sympathy as a tool to achieve the desired result.
C) it uses fear as a tool to achieve the desired result.

ANSWER

Answer: A

A – correct – Travis lavishes praise on his mother (flattery) and in conclusion states that therefore they should go to Disney World. But, a trip to Disney World is not justified because of his flattery alone. This flaw is called the "appeal to flattery" flaw.

B – incorrect. Travis is not appealing to anyone's sympathy.

C – incorrect – no reference to fear is present in the argument.

Answers
© Gift Of Logic, Inc * Copying prohibited

7 FLAWED REASONING

Traffic Cop: You made an illegal left turn ..
Car Driver: I make this left turn every day. If you watch..

The car driver's argument is fallacious because
A) it uses flattery to dissuade the traffic cop from issuing a ticket.
B) it attacks the character of the traffic cop to prove its point.
C) it appeals to common practice to justify its conclusion.

ANSWER

Answer: C

A – incorrect- the car driver does not flatter the traffic cop.

B – incorrect - the car driver does not call the traffic cop's character into question – he just argues with the cop.

C – correct - the car driver cites common practice by hundreds of other drivers to justify why he made the illegal left turn. This flaw is called the "appeal to common practice" flaw. Just because several others have made this illegal turn does not make it legal.

8 FLAWED REASONING

Tom: I don't like chocolate milk..
Dick: I love chocolate milk..
Harry: I also love chocolate..
Tom: Since all of you like chocolate milk..

Which one of the following most accurately describes the flaw in the Tom's line of reasoning?
A) he uses flattery to change his opinion on chocolates.
B) he appeals to the emotions of his friends.
C) he joins the bandwagon to avoid being singled out.

ANSWER

Answer: C

A – incorrect – no flattery can be identified in the argument.

B – incorrect – no emotions can be identified in the argument.

C – correct – Tom initially said that he does not like chocolate milk, but later decided that he also likes chocolate milk because his friends like them. Tom changed his opinion only to satisfy his friends. This reasoning is flawed and is referred to as "join the bandwagon" fallacy. Tom joined his friends' bandwagon in order to avoid being singled out.

9 FLAWED REASONING

Teacher: I am very excited. I will be teaching..
Student: Why are you excited? Will the new classroom..
Teacher: Since everything is new, you will score better than before.

The fallacy in the teacher's argument is that
A) it appeals to flattery to prove its point.
B) it appeals to novelty to prove its point.

ANSWER

Answer: B

A – incorrect – the teacher does not flatter the student.

B – correct – the teacher cites new furniture as the reason to conclude that they will score better. This flaw is called as the appeal to novelty flaw. It is not logical to expect new furniture to improve student scores.

| **10** | FLAWED REASONING |

He is a hardened criminal..

The fallacy in the argument is that
A) it provides irrelevant information to support the conclusion.
B) the conclusion does not take all the premises into consideration.

ANSWER

Answer: B

A – incorrect – the premises are relevant. Only the conclusion is flawed.

B – correct – the flaw in this argument is that not all evidence is considered in making the conclusion. Just because he did a few good things, he is being considered a good person. But, what about all the bad things that he did that landed him in jail? The conclusion does not take all the evidence into consideration. This flaw is called the flaw of insufficient evidence.

11	FLAWED REASONING

Andrei: I am throwing a birthday..
Sergei: Thanks for your invitation. I will surely..
Andrei: That's correct, I forgot that ..

The fallacy in the Andrei's argument is that
A) it appeals to authority to support its conclusion.
B) it appeals to spite to support its conclusion.

ANSWER

Answer: B

A – incorrect – there is no appeal made to anyone's authority.

B – correct – Sergei has petty ill will (spite) in his mind and influences Andrei not to invite Vladimir to the party. Andrei agrees to this spite and decides not to invite Vladimir to the party. This line of reasoning commits a fallacy of spite.

12 FLAWED REASONING

Nikki: I am going to team up with Cindy..
Vicki: You know, the last time I teamed up ..
Nikki: What you are saying is correct. They will not..

The fallacy in Nikki's argument is that
A) she changes her opinion because of the guilt of associating with Cindy and Sammy.
B) she is naive to trust Cindy and Sammy.

ANSWER

Answer: A

A – correct – Just because Cindy and Sammy did not work well in a project with Vicki does not mean that they will not work well with Nikki. This fallacy is called the guilt by association fallacy.

B – incorrect – Nikki does not trust Sammy and Cindy. Instead, Nikki thinks that Sammy and Cindy will take credit for her work.

13 FLAWED REASONING

Last year, two hundred students from the city..

The fallacy in the above argument is that
A) it assumes that what was true in the past will be true in the future.
B) it provides irrelevant data to support the conclusion.

ANSWER

Answer: A

A – correct – just because two hundred students attended last year is not sufficient evidence to conclude that two hundred students will attend this year as well. The flaw is in extrapolation from the past to the future.

B – incorrect – premises are not totally irrelevant for the argument.

14 FLAWED REASONING

The eggs in this shelf are broken..

The fallacy in the above argument is that
A) it takes a feature of one member of a group and applies it to the entire group.
B) it takes the feature of an entire group and applies it to one member of the group.

ANSWER

Answer: A

A – correct – the fallacy in this argument is that it takes a feature (broken eggs) of one member of the group (one shelf) and applies it to the entire group (all the shelves). This is a flaw of hasty generalization.
B – incorrect – this statement is an inaccurate description of the flaw.

15 FLAWED REASONING

Jane: This soccer team has great..
The fallacy in the above argument by Jane is that
A) it takes a characteristic of a group and applies it to one member.
B) it provides a conclusion that is in contradiction with its premises.

ANSWER

Answer: A

A – correct – the argument takes a group's (soccer team) characteristic (has several great players) and applies it to one member (Michael). There is a fallacy of division in this argument. It is possible that Michael is not a great player. B – incorrect – the conclusion is not in contradiction with the premises.

Answers
© Gift Of Logic, Inc * Copying prohibited

16 FLAWED REASONING

A frog jumps randomly from one..

The fallacy in the above argument is that
A) it improperly compares the random movement of a frog with the sequential movement of a postman.
B) it compares two similar movements and concludes that they are similar.

ANSWER

Answer: A

A – correct- the analogy between the frog and the postman is flawed. Frog moves randomly, whereas the postman moves sequentially.
B – incorrect- the random movement of a frog and the sequential movement of the postman are not similar.

17 FLAWED REASONING

Jason graduated high school with good academic..

The fallacy in the above argument is that
A) it assumes that he does not have any other choice except the Engineering and Medical professions.
B) the premise is totally irrelevant to the conclusion.

ANSWER

Answer: A

A – correct – this is a fallacy of choice. There are other professions that he can choose such as Arts, Pharmacy, etc.
B – incorrect. The premise is not totally irrelevant.

18 FLAWED REASONING

Nearly one hundred people gathered at the Innovation..

The fallacy in the above argument is that
A) it fails to consider the fact that Mr. Geekman has a good reputation in the community.
B) is assumes that if a claim is not proved once, it is a false claim.

ANSWER

Answer: B

A – incorrect - Mr. Geekman's reputation in the community is not stated as a premise in the argument. So, this is not the correct choice.

B – correct – Mr. Geekman claimed that the robot can cut vegetables. But, he failed to prove his claim because the robot's hand broke during the demonstration. So, to conclude that Mr. Geekman made a false claim is not correct reasoning. Such a reasoning assumes that if a claim cannot be proved once, it is a false claim. Perhaps he could do another demonstration to prove his claim.

19	FLAWED REASONING

The Greenleaf party is certain that they will win..

The fallacy in the above argument is that
A) it takes the results from an unbiased sample and applies it to the entire voting population.
B) it takes the results from a biased sample and applies it to the entire voting population.

ANSWER

Answer: B

A – incorrect – the sample that is used for the survey is biased because it is made up entirely of the party's supporters.

B – correct – the survey was done on the party's supporters, but the conclusion was applied to the entire voting population. This fallacy is called the fallacy of biased sample.

20 FLAWED REASONING

The principal of a high school was tasked with selecting..

The fallacy in the principal's method of selecting the team is that
A) it does not provide a rationale to justify its decision.
B) it predetermined the composition of the sample.

ANSWER

Answer: B

A – incorrect – it provided a rationale as to why she picked 2 boys and 4 girls – because in previous years, this combination of 2 boys and 4 girls had won medals for the school. But, the principal was tasked with making an unbiased selection.

B – correct – the principal was tasked with picking the team without bias, but she was biased in her selection. She picked the team with the same number of boys and girls that were in the winning team during the previous years. This is a fallacy of predetermination/bias. She predetermined the composition of the team.

Answers
© Gift Of Logic, Inc * Copying prohibited

21 FLAWED REASONING

Inspector: Last month, I ordered you to upgrade..
Restaurant Owner: Here are the receipts showing..
Inspector: I am not satisfied..

The fallacy in the Inspector's decision to impose a fine is that
A) it attacks the owner's character.
B) it willfully ignores the evidence presented.

ANSWER

Answer: B

A – incorrect – there is no attack on the owner's character by the inspector.

B – correct – the inspector ignores the receipts that were produced before him proving that the upgrades were completed. This flaw is called the flaw of willful ignorance.

22 FLAWED REASONING

Sabrina: The new "Zoro" hybrid car in the market..
Mariana: Myself and my father love this car very much..
Sabrina: I love the features ..

The fallacy in Sabrina's argument is that
A) it appeals to popular appeal to reach a conclusion.
B) it ignores unfavorable evidence and considers only the favorable evidence.

ANSWER

Answer: B

A – incorrect - Mariana's opinion alone does not make it a popular opinion. No facts are provided to indicate that this is a popular opinion.

B – correct – Sabrina picks only the facts that are favorable and ignores the unfavorable facts. This kind of fallacy is called the cherry picking fallacy.

23 FLAWED REASONING

Several teams of police detectives were hot in trail..

The fallacy in the above argument is that
A) it jumps to a conclusion with little or no evidence.
B) it bases its decisions on irrelevant pieces of information.

ANSWER

Answer: A

A – correct – there is no evidence to suggest that the man that the police were pursuing is the exact same man mentioned in the television. The police have closed the case without checking this fact. This type of fallacy is called the "jumping to conclusion" fallacy.

B – incorrect – the premises are relevant, but not sufficient to infer the conclusion.

24 FLAWED REASONING

To reduce overcrowding in trains..

The fallacy in the above argument is that
A) it appeals to emotion rather than providing facts to support its conclusion.
B) it appeals to majority's opinion to make its conclusion.

ANSWER

Answer: B

A – incorrect – no appeal to emotion is made in this argument.

B – correct – just because ninety percent of the commuters think that more trains must be introduced to reduce overcrowding does not mean that they are correct. There may be better ways to reduce overcrowding. The fallacy in this argument is that it appeals to majority's opinion. It is called as the appeal to majority fallacy.

25 FLAWED REASONING

Newspaper Article: We recommend Mr. Harry.

Which one of the following most accurately describes a flaw in the arguments reasoning?
A) it appeals to the pity of the readers to win support for its conclusion.
B) it resorts to cherry-picking to prove its conclusion.

ANSWER

Answer: A

A – correct – it appeals to the readers in a tone that seeks the readers to take pity on Mr. Harry and vote for him. This flaw is called the "appeal to pity/sympathy" flaw.

B – incorrect – cherry-picking exists when only the favorable facts are selected to reach the conclusion and unfavorable facts are ignored. There is no cherry-picking involved in this argument.

26 FLAWED REASONING

The Y-series of the Zoro line of hybrid cars..

The fallacy in the above argument is that
A) the premises in the argument contradict the conclusion.
B) it appeals to envy to reach its conclusion.

ANSWER

Answer: A

A – correct – the premise is that these cars will be used by very affluent people, but the conclusion is that it is very affordable, which means that one does not have to be very affluent to buy these cars. The conclusion cannot be derived from the premises. They are in contradiction with each other.

B – incorrect – if it appealed to envy, then it is not likely to be very affordable.

27 FLAWED REASONING

Pam: I want to give the Biggest Eater award..
Sam: Easily! I can eat a hundred pancakes..
Pam: Ok. I hereby confer upon you..

The fallacy in Pam's argument is that
A) it does not provide a rationale to justify its decision.
B) it relies on hyperbole to infer its conclusion.

ANSWER

Answer: A

A – incorrect – when Pam says "ok", she agrees to Sam's claim that he can eat a lot.

B – correct – hyperbole means exaggerated claim. Pam made a decision to award Sam with the title just based on his hyperbolic claim, without seeing any other proof or giving others a chance to contest for the award.

1 ANALOGOUS REASONING

All the cars, when considered together..

The flawed pattern of reasoning in the argument above is similar to that in which one of the following?
A) Joan loves democracy. So, the people of her country also love democracy.
B) All the fruits are healthy to eat. So, the apples are healthy to eat.
C) All the trucks, when considered together, pollute the air very much. So, Nancy's truck also pollutes the air very much.

ANSWER

Answer: C

    Premise: All the cars, when considered together, use a lot of iron.
    Conclusion: So, Stanley's car also uses a lot of iron.

The premise is about cars as a whole, but the flawed conclusion is about one particular car. We need to find an answer choice that has this same type of flaw in its reasoning. Note the question gives a hint that the argument is flawed.

A – incorrect – this argument is opposite in its reasoning - it takes one person's desire and assumes that her entire country desires the same.

B – incorrect – The conclusion of this argument is not about one apple, but it is about several apples. This is not analogous to the original argument.

C – correct - this argument has the same type of premise and conclusion as the given argument.

2 ANALOGOUS REASONING

If a person plays the piano, that person..
The reasoning in the argument above is parallel in its structure to which one of the following?
A) If a person drives a car, that person must be a good driver. Sam does not drive a car and so he must not be a good driver.
B) If a person draws well, that person must be creative. Todd draws well and consequently he must be creative.

ANSWER

Answer: B

Note the conditional premise in the argument.
    person plays the piano → person must be talented
    Tina plays the piano.
    Therefore, she must be talented.

A – incorrect
    person drives a car → person must be a good driver
    Sam does not drive a car. (not analogous)
    So, he must not be a good driver. (not analogous)

B - correct - the premise and conclusion of this argument is analogous to the original argument.

    person draws well → person must be creative
    Todd draws well.
    Consequently, he must be creative.

3 ANALOGOUS REASONING

Ian's mom told him that if he studies..

The flawed pattern of reasoning in the argument above is similar to that in which one of the following?

A) If the train maintains its speed, it will arrive at the station in time. The train arrived in time; so it must have maintained its speed.

B) If it snows, then the temperature will be cold. So, since it did not snow, the temperature was not cold.

ANSWER

Answer: A

    Ian studies for two hours → he can eat an ice-cream

    Ian ate an ice-cream.

    Therefore, he studied for two hours.

The flaw in this argument is that it interprets the condition incorrectly. Just because he ate an ice-cream today does not imply that he studied for two hours. The conclusion is the converse of a conditional.

A - correct - this argument has a parallel conditional premise and an incorrect converse conclusion similar to the main argument.

    train maintains its speed → will arrive in time

    Train arrived in time.

    Therefore it maintained its speed. (converse of the conditional)

B - incorrect - this argument has a different structure than the original. The premise is conditional, but the conclusion is the inverse of the conditional.

    it snows → temperature will be cold

    it did not snow.

    So, the temperature was not cold.

4 ANALOGOUS REASONING

Before going to the backyard, Vivian..

The reasoning in the argument above is analogous to which one of the following?
A) Ronald was alerted by his mom not to get wet in the rain since he could damage his watch. But, Ronald reasoned that his watch was water proof. Therefore, he dismissed the warning.
B) Amber's mom cautioned her to be careful while handling the electric drill. Amber argued that the new drill was safe. So, she ignored the warning to be cautious.

ANSWER

Answer: A

    Vivian was asked to be careful about fire ants.
    Vivian claimed that his home was new and will not have fire ants.
    So, he ignored the warning.

A - incorrect - Ronald argued that his watch was water proof, but did not say that his watch was new. This premise is not analogous.

B - correct - this choice is analogous to the original argument

    Amber was cautioned to be careful while using the drill.
    Amber said that the drill was new and safe.
    So, she ignored the warning.

5 ANALOGOUS REASONING

All P are Q. All Q are R. Therefore, all P are R.

Which one of the following arguments has a reasoning similar to the argument above?

A)
All M are N.
If there is a N, then it is a O.
Therefore, all O are M.

B)
All A are B.
If there is a B, then it is a C.
Therefore, all A are C.

ANSWER

Answer: B

    All P are Q.   All Q are R.   Therefore, all P are R.

A - The conclusion in this argument is not analogous.
    All M are N.
    All N are O (same as If it is a N then it is a O)
    Therefore, All O are M. (this should have been all M are O to be analogous)

B - correct - the premises and conclusion are analogous to the original.
    All A are B.
    All B are C. (same as If it is a B, then it is a C)
    Therefore, all A are C.

6 ANALOGOUS REASONING

A doctor attending to a patient noticed..

The flawed pattern of reasoning in the argument above is similar to that in which one of the following?

A) The home owner noticed that when he walked up the stairs, the stairs swayed first, and then there was a squeaking noise. So, he concluded that the squeaking noise made the stairs to sway.

B) When Lucy was driving her car, she first noticed a blue van and then after a minute, she witnessed an accident. So, she concluded that the blue van caused the accident.

ANSWER

Answer: B

Note the causality in the argument.
  high temperature noticed first, then high blood pressure
  therefore, high temperature c→ blood pressure

A - incorrect - the conclusion in this argument has its cause and effect reversed.
  stairs swayed first and then squeaking noise
  therefore, squeaking noise c→ swaying stairs

B - correct - the premise and conclusion are analogous.
  Lucy noticed blue van first and then an accident
  therefore, blue van c→ accident

The symbol c→ refers to the causal relationship. Consult the book titled "Critical thinking & Logical reasoning Primer" for a detailed discussion.

7 ANALOGOUS REASONING

P, Q, and R are events that happened at different times...

Which one of the following arguments has a reasoning similar to the argument above?

A) L, M, and N are events that happened at different times. L caused M to happen. M triggered N. Therefore, L is the effect of N.

B) I, J, and K are events that happened at different times. J is the effect of I and J is the cause of K. Therefore, K is the effect of I.

ANSWER

Answer: B

The given argument tests your ability to understand causal relationships. Drawing a causal diagram will help map the analogies.

$P \xrightarrow{c} Q$
$Q \xrightarrow{c} R$
Therefore, $P \xrightarrow{c} R$

A - incorrect - the conclusion is not analogous to the original conclusion.

$L \xrightarrow{c} M$
$M \xrightarrow{c} N$
$N \xrightarrow{c} L$ (this conclusion should have been reversed to be analogous)

B - correct - the premises and conclusion are analogous.

$I \xrightarrow{c} J$
$J \xrightarrow{c} K$
Therefore, $I \xrightarrow{c} K$

8 ANALOGOUS REASONING

Of late, incidences of theft of watches have increased...

Which one of the following arguments has a reasoning similar to the argument above?

A) In recent days, cars are being stolen in record numbers. Consequently, anyone must be able to prove ownership of their vehicles when asked to do so.

B) Lots of computers are being stolen these days. Therefore, people must show proof of ownership of their computers on request.

ANSWER

Answer: B

> theft of watches is increasing.
> So, people must prove ownership of watches when requested.

Note carefully that this argument concludes that ownership of "watches" must be proved because "watches" are being stolen.

A - incorrect

> cars are being stolen in record numbers
> So, prove ownership of vehicles when requested.

The conclusion does not refer to "cars" but instead refers to all vehicles. This is not the same type of reasoning found in the original argument.

B - correct - this argument is analogous to the original argument.

> lots of computers are being stolen.
> So, people must prove ownership of their computers on request.

Answers

9 ANALOGOUS REASONING

Rhonda was sure that facial cream..

The pattern of reasoning in the argument above is similar to which one of the following?

A) Alex was sure that the ant spray that he used will kill the ants in the kitchen. But, even after spraying a lot the ants did not die. So, Alex blamed the ants for the problem.

B) Bruce kept wiping the table hard with a grease remover because he wanted to get rid of the sticky substance on the table. The sticky substance did not go away. So, Bruce blamed the grease remover for the problem.

ANSWER

Answer: A
- facial cream was applied to remove pimples
- but, pimples did not go away
- So, pimples were blamed for the problem

This argument blames the pimples instead of the facial cream.

A - correct - This argument is analogous to the original argument since it blames the ants instead of the ant spray.
- spray was used to kill the ants in the kitchen
- but, the ants did not die
- So, the ants were blamed for the problem

B - incorrect - If this argument is to be analogous to the original argument, then the sticky substance must be blamed.
- the grease remover was used to remove the sticky substance
- but, sticky substance did not go away
- So, the grease remover was blamed

10 ANALOGOUS REASONING

Most people like to lead peaceful lives..

The reasoning in the argument above is analogous to which one of the following?
A) Most children like to eat healthy food. So, food items that are spicy must not be served to them.
B) Most people like humor. So, television programs that show serious content must not be telecast.

ANSWER

Answer: A

    Most people like to lead peaceful lives.
    So, radio programs dealing with controversies must not be aired.

The reasoning in this argument is that peaceful lives and controversies do not go together. We need to look for this type of analogy in the choices.

A - incorrect
    Most children like healthy food.
    So, spicy food items must not be served.
Healthy food and spicy food could go together. So, this argument is not analogous to the original argument.

B - correct
    Most people like humor.
    So, television programs that show serious content must not be aired.
Humor and serious content do not go together. This argument is analogous to the original argument.

Answers
© Gift Of Logic, Inc * Copying prohibited

11 ANALOGOUS REASONING

It is painful to get a bank loan..

Which one of the following arguments has a reasoning similar to the argument above?

A) It is a hassle to clear the security at the airport. They check a lot of things. So, the security process must be modified.

B) It is difficult to rent a car. You never get the car that you want. So, the car renting process must be revised.

ANSWER

Answer: A

A – correct - this argument is analogous to the original argument. The analogies are shown below.

    Painful to get loan ↔ hassle to clear airport security
    Ask a lot of information ↔ check a lot of things
    So, process must be revamped ↔ So, process must be modified

B – incorrect - "asking for a lot of information" and "not getting the car that you want" are not analogous.
    Painful to get loan ↔ difficult to rent a car
    Ask a lot of information ↔ you never get the car that you want
    So, process must be revamped ↔ So, process must be revised

Answers
© Gift Of Logic, Inc * Copying prohibited

12 ANALOGOUS REASONING

The pilot of an airplane decided to make an emergency..

The flawed pattern of reasoning in the argument above is similar to that in which one of the following?

A) A man in an air balloon requested help to land. So, the helpers on the ground decided that he was not feeling well.

B) The helicopter pilot requested help to land immediately. Therefore, the traffic controllers decided that there was some problem with the helicopter.

ANSWER

Answer: B

Airplane pilot requested help for emergency landing.
So, the controllers assumed technical difficulties with the plane.

A - incorrect
airplane pilot requested help ↔ man in balloon requested help
So, the controllers assumed technical difficulties with the plane. ↔ So, helpers decided that he was not feeling well

This is not analogous to the original argument. To be analogous, there should have been some technical difficulty with the balloon.

B – correct
airplane pilot requested help ↔ helicopter pilot requested help
So, the controllers assumed technical difficulties with the airplane ↔ So, controllers decided that there was a problem with the helicopter

This argument is analogous to the original argument. The traffic controllers decided that there was some problem with the helicopter, but not with the helicopter pilot.

13 ANALOGOUS REASONING

Three dress sizes, small, medium and large..

The pattern of reasoning in the argument above is similar to that in which one of the following?

A) Gloves are available in two sizes: small and medium. These sizes may not fit everyone. So, some must hire a glove maker to make one that fits.

B) One has three choices of jeans – straight, pleated and relaxed. One may not find these sizes suitable. So, they must wear the one that fits the best.

ANSWER

Answer: A

    three dress sizes are available - small, medium and large
    but, not everyone can fit into these sizes
    So, some must hire a tailor to make a dress that fits

A - correct

    gloves are available in two sizes - small and medium
    sizes may not fit everyone
    So, some must hire a glove maker to make one that fits

The argument that if gloves do not fit, then they should hire a glove maker is analogous to the original argument. Even though there are only two types of gloves whereas there are three types of dresses, the substance of the analogy is conveyed by this argument.

B - incorrect

    jeans in three types- straight, pleated and relaxed.
    one may find these sizes unsuitable
    So, they must wear the one that fits the best

This argument does not recommend a jeans maker to make a jeans that fits. This is not analogous to the original argument.

14 ANALOGOUS REASONING

My first car was made by Honda..

The flawed pattern of reasoning in the argument above is similar to that in which one of the following?

A) My first watch was made by Timex. My second watch was also made by Timex. My first watch lasted for fifteen years. Therefore, my second watch will also last for fifteen years.

B) My first pet was a dog. My second pet was also a dog. My first pet lived for ten years. Therefore, my second pet will live for twenty years.

ANSWER

Answer: A

The argument is that both the first and second cars were made by Honda and since the first car lasted twenty years, it concludes that the second car also will last twenty years. This argument contains an analogy within itself (analogy between the two cars).

A - correct - this argument is analogous to the original argument. The first and second watches were made by the same company, Timex. Since the first watch lasted for fifteen years, the second watch will also last for fifteen years. Note that there is an analogy within this argument itself (analogy between the two watches).

B - incorrect - this is not analogous to the original argument. The first pet lived for ten years and the second pet is also expected to live for ten years. But, the second pet lived for twenty years. Note that this argument has an analogy within itself (analogy between the two dogs).

| 15 | ANALOGOUS REASONING |

Jack loves cooking..

The pattern of reasoning in the argument above is similar to that in which one of the following?

A) Martha likes music. Martha and Maggie have a sister in common. So, Maggie also likes music.

B) Rick likes Science. Rick and Sam have a brother in common. So, Sam likes History.

ANSWER

Answer: A

    Jack loves cooking.
    Jack and Jill have the same parents.
    So, Jill also loves cooking.

Jack and Jill both have the same parents and both love cooking.

A - correct - the arguments are analogous.
    Martha likes music.
    Martha and Maggie have a sister in common.
    So, Maggie also likes music.

Martha and Maggie both have a sister in common and both love music.

B - incorrect
    Rick likes Science.
    Rick and Sam have a brother in common.
    So, Sam likes History.

Rick and Sam both have a brother in common, but only one likes Science.

Answers

1 SUDOKU

Solve the following Sudoku. A correctly solved Sudoku has numbers 1-9 appearing only once in each row, each column and each 3x3 grid. Solving Sudokus will help you to gain valuable analytic skills.

4	6	3	7	2	5	8	1	9
5	8	9	4	1	3	7	6	2
1	2	7	9	8	6	5	3	4
6	9	1	2	4	8	3	7	5
7	4	5	3	6	1	2	9	8
2	3	8	5	7	9	1	4	6
9	7	6	1	5	2	4	8	3
8	1	2	6	3	4	9	5	7
3	5	4	8	9	7	6	2	1

Answers
© Gift Of Logic, Inc * Copying prohibited

2 SUDOKU

Solve the following Sudoku. A correctly solved Sudoku has numbers 1-9 appearing only once in each row, each column and each 3x3 grid. Solving Sudokus will help you to gain valuable analytic skills.

4	3	1	2	9	5	6	8	7
9	8	7	4	1	6	3	5	2
5	6	2	3	8	7	9	4	1
8	7	3	5	2	4	1	9	6
6	2	9	1	7	8	5	3	5
1	4	5	9	6	3	2	7	8
2	9	4	7	5	1	8	6	3
3	5	8	6	4	2	7	1	9
7	1	6	8	3	9	5	2	4

Answers
© Gift Of Logic, Inc * Copying prohibited

3   SUDOKU

Solve the following Sudoku. A correctly solved Sudoku has numbers 1-9 appearing only once in each row, each column and each 3x3 grid. Solving Sudokus will help you to gain valuable analytic skills.

9	6	7	8	1	2	5	4	3
4	8	3	6	5	9	1	2	7
2	1	5	4	3	7	8	9	6
7	9	8	1	4	5	3	6	2
5	3	4	2	8	6	9	7	1
1	2	6	7	9	3	4	5	8
8	4	2	9	6	1	7	3	5
6	5	1	3	7	4	2	8	9
3	7	9	5	2	8	6	1	4

4   SUDOKU

Solve the following Sudoku. A correctly solved Sudoku has numbers 1-9 appearing only once in each row, each column and each 3x3 grid. Solving Sudokus will help you to gain valuable analytic skills.

5	6	2	7	8	1	9	3	4
1	3	7	9	4	6	2	5	8
9	4	8	5	2	3	7	6	1
7	1	4	8	6	5	3	9	2
8	2	5	1	3	9	4	7	6
3	9	6	4	7	2	8	1	5
2	7	3	6	5	8	1	4	9
6	8	9	3	1	4	5	2	7
4	5	1	2	9	7	6	8	3

Answers
© **Gift Of Logic, Inc** * **Copying prohibited**

5 SUDOKU

Solve the following Sudoku. A correctly solved Sudoku has numbers 1-9 appearing only once in each row, each column and each 3x3 grid. Solving Sudokus will help you to gain valuable analytic skills.

7	8	4	2	5	9	6	1	3
6	3	5	7	1	8	9	2	4
9	1	2	6	4	3	5	7	8
5	6	9	8	2	4	1	3	7
1	4	8	3	7	6	2	9	5
2	7	3	5	9	1	8	4	6
3	5	1	4	6	2	7	8	9
4	2	7	9	8	5	3	6	1
8	9	6	1	3	7	4	5	2

answers

1 POSITIONING

Three boys Andrew, Bashir, and Chen..
Note that there are three boys that need to be seated in four chairs. So, one chair will be vacant.
Andrew must sit immediately to the left of Bashir >> AB

1) If the fourth chair must be left vacant, list all the possible ways in which the three boys can be seated?

A,B,C
AB

1	2	3	4
A	B	C	X
C	A	B	X

Answer: A,B,C,X and C,A,B,X, where X refers to a vacant chair.

2) If the third chair must be left vacant, then which of the following must be true? Answer: C- Chen must sit in the fourth chair. See the diagram below.

A,B,C
AB

1	2	3	4
A	B	X	C

3) If the second chair is not occupied, then which of the following is possible? Answer B-Chen, Vacant, Andrew, Bashir. See the diagram below.

A,B,C
AB

1	2	3	4
C	X	A	B

Answers
© Gift Of Logic, Inc * Copying prohibited

2 POSITIONING

Three girls Anita, Beena, and Cathy..

Note that 3 people must sit in 4 chairs, thus creating one vacancy.

Someone must sit in chairs 2 and 3 means that chairs 2 and 3 cannot be vacant >> ~X2, ~X3, where X represents vacancy.

Anita must be immediately to the left of Beena >> AB

1) Which of the following seatings are not possible?
Answer: A, D

A,B,C
AB
~X2, ~X3

1	2	3	4
A	B	X	C
A	B	C	X
X	A	B	C
C	X	A	B

Apply the rules to each of the following choices.

    A)   Anita, Beena, Vacant, Cathy >> not possible - violates ~X3 rule
    B)   Anita, Beena, Cathy, Vacant >> possible
    C)   Vacant, Anita, Beena, Cathy >> possible
    D)   Cathy, Vacant, Anita, Beena >> not possible - violates ~X2 rule

Answers
© Gift Of Logic, Inc * Copying prohibited

3 POSITIONING

3 girls Anita, Beena, and Cathy.. >> A,B,C

Someone must sit in chairs 2 or 3. This is an interesting rule. Logically, chair 2 may be occupied and chair 3 may be vacant or, chair 3 may be occupied and chair 2 may be vacant. This can be represented as X2 ╫ X3. Both chairs 2 and 3 cannot be vacant as the rule states that someone should sit in either chairs. Note that this rule allows both chairs to be occupied. The rules do not say that only chairs 2 or 3 can be vacant.

Chairs 2 or 3 can be vacant >> X2 ╫ X3
Anita must sit immediately to the left of Beena. AB

1) Which of the following seatings are possible?

A,B,C
AB
X2 ╫ X3

1	2	3	4
A	B	X	C
X	C	A	B
X	B	A	C
C	X	A	B

Answer: A, B, D
  A) Anita, Beena, Vacant, Possible >> possible
  B) Vacant, Cathy, Anita, Beena >> possible.
  C) Vacant, Beena, Anita, Cathy >> not possible - violates rule AB
  D) Cathy, Vacant, Anita, Beena >> possible

Answers
© Gift Of Logic, Inc * Copying prohibited

4 POSITIONING

There are 4 branches and three birds. So, one of the branches will be vacant.

1) If B and C must sit in the right side of the tree, and C must sit below B, then A must sit opposite to B.

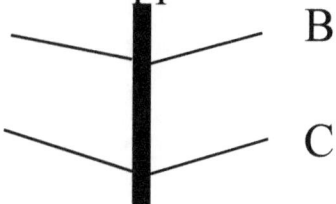

Answer: B) False After diagramming the rules, it is clear that A need not sit opposite to B. A can sit opposite to C.

2) If C and B can sit on either side of the tree, and C must be below B, then which one of the following must be true?

Answer: A) Any branch can be vacant. See diagram below where X represents vacancy.

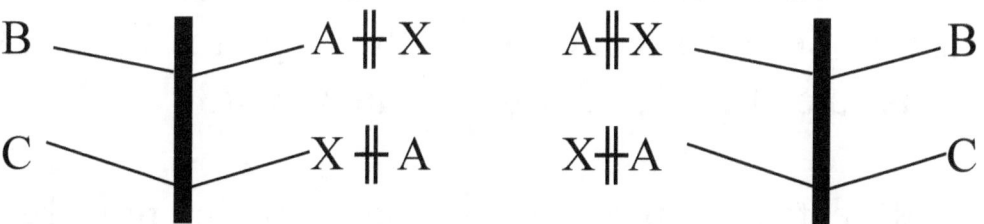

If C and B are on the left side, then depending on where A sits on the right side, either of the positions on the right side can be vacant. Similarly, if C and B are on the right side, depending on where A sits on the left side, either of the positions on the left side can be vacant. So, choice A must be true. Choice B - Bird A must always sit on a top branch is clearly not true always. A ⫲ X means A or Vacant (X).

Answers

5 POSITIONING

There are four types of cheese in the kitchen..
Diagram the scenario as follows. Note that there are three rats, but four types of cheese. So, one type of cheese will not be eaten.

R1 can eat C1 or C3 >> R1C1╪R1C3
R2 can eat C2 or C4 >> R2C2╪R2C4
R3 can eat C3 or C1 >> R3C3╪R3C1

R1,R2,R3
R1C1╪R1C3
R2C2╪R2C4
R3C3╪R3C1

C1	C2	C3	C4
R1	R2	R3	X
R1	X	R3	R2

1) If R1 eats C1 and R2 eats C2 then which of the following must be true?
Answer: C. R3 eats C3. See the second row in the diagram. Since R3 can eat C3 or C1 and C1 is eaten by R1, C3 will be eaten by R3.

2) If R1 eats C1 and R2 eats C4, then which of the following must be true?
Answer: A) R3 eats C3 and C2 is not eaten. See the third row in the diagram.

In this problem, the Rats are "positioning" themselves to eat the cheese according to various constraints. Positioning problems do not always have to position people in chairs. You also encounter them when you park your car in a parking lot or fix appointments on a calendar.

Answers
© Gift Of Logic, Inc * Copying prohibited

6 POSITIONING

A doctor has four appointments ..
Note that there are 3 patients and 4 appointments available. This means one of the appointments will not be filled. Represent the rule that at least one morning and one evening must be filed as >> 1 AM, 1 PM to help remember it as you solve the problem.

1) The number of ways appointments can be scheduled so that patient A is seen first in the morning and patient B is also seen in the morning is
Answer: B) 2. Diagram the problem as shown below. Patient C can be seen in either the first or the second afternoon appointments.

1AM, 1PM

1	2	3	4
A	B	C	X
A	B	X	C

2) The number of ways appointments can be scheduled so that only patient B is seen in the afternoon is
Answer: C) 4 . See diagram below. If only patient B is seen in the afternoon, then patient A and patient C have to be seen in the morning. The diagram below shows all the possibilities.

1AM, 1PM

1	2	3	4
A	C	B	X
A	C	X	B
C	A	B	X
C	A	X	B

Answers
© Gift Of Logic, Inc * Copying prohibited

6   POSITIONING

3) The number of scheduling possibilities in which only patient A can meet the doctor in the morning and patient B is the first to see the doctor in the afternoon is

Answer: A) 2. See diagram below which shows the possibilities.

1AM, 1PM

1	2	3	4
A	X	B	C
X	A	B	C

Answers
© Gift Of Logic, Inc * Copying prohibited

7 POSITIONING

Four sailors A, B, C, and D..

Note that there are 4 sailors, but only 3 boats. So, one sailor will be unassigned. If C is assigned to sail, D must also sail >> C → D

1) If A sails, then which sailor will not be able to sail?
Answer: B) B or C. See the diagram below. If A sails and C sails, then D must sail leaving B out. If A sails and B sails, then D can sail leaving C out. The diagram below shows the possibilities where B or C is left out.

	1	2	3
A,B,C,D	A	B	D
C → D	A	C	D

2) If D does not sail, then which of the following will not sail?
Answer: C) C. This question actually tests your understanding of the contrapositive.

    C → D   if C sails, D sails
    ~D → ~C if D does not sail then C does not sail
            (contrapositive of C → D)

3) If D does not sail, then the number of unassigned boats will be
Answer: B) 1

If D does not sail, then C cannot sail per contrapositive discussed above. So, only A and B can sail in two boats, leaving one boat unassigned.

Answers
© Gift Of Logic, Inc * Copying prohibited

8 POSITIONING

Four patients, A, B, C, and D need appointments..
Note that there are 4 patients, but only 3 appointments available. So, one patient has to wait until the next week. M-Monday, T-Tuesday

```
   M     T
-  -     -
```

1) If one appointment on Monday is filled by A and the appointment on Tuesday is filled by C, then which of the following must be true?
Answer: B) Exactly one among B or D will not be able to see the doctor this week.

```
   M     T
-  -     -
A‖ A‖   C
```

Answer: C is seen on Tuesday and A on Monday. This means, that there is one appointment available for B or D. So, one among them will not be able to see the doctor this week. A‖ means, A can be at this position or in another position.

2) If A and C must be seen on the same day, then which of the following must be true?
Answer: B) D cannot be seen on the same day as C.

If A and C must be seen on the same day, it has to be on Monday, since two appointments are available only on Monday. So, B or D can only be seen on Tuesday. D cannot be seen on the same day as C which is a Monday.

```
   M     T
-  -     -
A  C    B‖D
```

Answers   130
© Gift Of Logic, Inc * Copying prohibited

9 POSITIONING

The availability of GreatLife recreation center..

	Monday	Tuesday	Wednesday	Thursday	Friday
morning	open	open	open	open	open
afternoon	open	closed	open	open	closed

The rule can be represented as M(today) → A (next), where M is for morning and A is for afternoon.

1) Can he play on Monday morning?

Answer: B) No. Looking at the chart and keeping the rule in mind, we can see that if he plays on Monday morning, he must play on Tuesday afternoon. But, the recreation center is closed on Tuesday afternoon, so he cannot play on Monday morning.

2) Can he play on Friday morning?

Answer: A) Yes. If he plays on Friday morning, then he can play on Monday afternoon, since the recreation center is open then.

3) How many mornings can he play in a week?

Answer: B) 3. He can play on Tuesday, Wednesday and Friday morning only since only Wednesday, Thursday and Monday afternoons are open.

answers
© Gift Of Logic, Inc * Copying prohibited

10   POSITIONING

Four people A, B, C, and D..

A must sit next to B on the same row >> AB ‖ BA
B must sit above C >> B
                                    C

D must sit immediately after C on the same row >> CD

We can chain the rules together as follows. Note that A can be before B or after B. So, we have two possibilities as shown below.

1)   A B   2)   B A
     C D   C D

1) If B sits in the top row-second position, then there is one vacant spot in each row. Answer: A) True.

1	2	3
A‖X	B	A‖X
X	C	D

2) If C sits in the bottom row-first position, then the vacant positions are one below the other. Answer: A) True. See the diagram below.

1	2	3
B	A	X
C	D	X

3) If B sits in the top row-second position and the spot immediately to the left of B is vacant, then D sits in the second row-second position. Answer: A) False. D sits in the second row-third position.

1	2	3
X	B	A
X	C	D

Answers
© Gift Of Logic, Inc * Copying prohibited

11 POSITIONING

The postman to Dove Creek Lane..

7 packages, 3 homes.
At least one package to each home.
Packages must be delivered in descending order.

1) If he delivers one package to A and four packages to C, then the order of his trip to the three homes is
Answer: A) C, B, A. See diagram below. If he delivers 1 package to A and 4 packages to C, he will deliver the remaining 2 packages to B. So, the order in which he will deliver in descending order is C,B,A.

A	B	C
1	2	4

2) If he delivers to B first and could deliver next to either A or C, then the number of packages he delivers to A,B,C is
Answer: B) A-1, B-5, C-1

Remember that he must deliver in descending order of the number of packages to each home. The question states that he delivers to B first. So, B must have the maximum number of packages. It also states that he could deliver next to either A or C. This means that A and C have both the same number of packages. Only answer choice A, namely A-1,B-5,C-1 has the maximum number of packages for B, has the same number of packages for A and C, and a total of seven packages.

Answers
© Gift Of Logic, Inc * Copying prohibited

1 GROUPING AND POSITIONING

Three flowers are to be selected for planting from a group of four..
   Red, Green, Blue, Yellow >> R,G,B,Y

If blue is selected, then it must be planted after red >> B → R-B. Note that R-B represents a rule regarding the position of the flowers, whereas the B→ is regarding the selection of the flowers. B→ means "if B is selected then". So, interestingly, this rule represents a combination of a selection rule and a positioning rule. Also, B → R-B means that R must be selected if B is selected.

1) Which of the following choices satisfy the selection and positioning constraints for planting in the garden?

Answer: B, C, D

Apply the rule B → R-B to all the choices. Note that if B is selected, then R must be selected and also R must be planted before B. Look for BR to satisfy the selection criteria, and look for R-B to satisfy the positioning criteria.

A)	Blue, Red, Yellow	invalid, B is selected, but it is planted after R.
B)	Red, Blue, Yellow	valid, B is selected and is planted after R.
C)	Red, Green, Blue	valid, B is selected, and is planted after R.
D)	Green, Yellow, Red	valid, B is not selected, but that is ok.

Answers
© Gift Of Logic, Inc * Copying prohibited

2   GROUPING AND POSITIONING

Tracy plans to invite only three families for dinner..

Family F2 can be invited on Wednesday only >> F2@Wed
Note that this rule does not mean that only F2 can be invited on Wednesday. Some other family can be invited on Wednesday.

1) Which of the following invitations can be made for Monday, Tuesday, and Wednesday?

Answer: A, C

	M	T	W	
A)	F1	F3	F2	>> correct, invitation can be made.
B)	F2	F2	F4	>> incorrect, F2 can not be invited on Tuesday.
C)	F1	F3	F4	>> correct, invitation can be made without F2.

2) If F2 must be invited, and F1 is also invited, then which one of the following families can also be invited?

Answer A) F3, F4.

If F1 and F2 are invited, there will be one other family that can be invited. The remaining ones are F3 and F4. One of these families can be invited.

3   GROUPING AND POSITIONING

Sheela is moving into an apartment and is setting..

T, S, C, P represents TV, Sofa, Computer, and Printer respectively.

The computer and printer must be in the same room. This means that they must be together. This can be represented a CP.

If the TV is placed in a room, the Sofa also must be placed in the same room. This can be represented as T→S.

If the Sofa is placed in a room, the computer also must be placed in the same room. This can be represented as S→C.

We can chain the rules as T→S →CP. Remember that the arrow → represents conditionality. So, T→S does not imply S→T. (converse of a conditional is not true).

1) Which of the following assignments of items is correct?

Answer: C.

    A) TV, Sofa, Computer >> incorrect , Printer is missing
    B) TV, Computer, Printer >> incorrect, Sofa is missing
    C) Sofa, Computer, Printer >> correct, satisfies S → CP

Answers
© Gift Of Logic, Inc * Copying prohibited

4 GROUPING AND POSITIONING

Three statues must be picked out from the available..

The red and blue statues must be selected >> RB

If the blue statue is placed in the third position, the green statue must be placed in the first position. >> B3 → Y1

1) Which of the following combinations of statues is a valid selection? Note that this question does not mention about the positions of the statues. It only asks if the given selections are valid or not. So, the correct answers must satisfy the RB rule only.
   A) Red, Green, Blue >> valid
   B) Green, Blue, Yellow >> invalid, violates RB rule
   C) Blue, Yellow, Red >> valid
   D) Yellow, Red, Green >> invalid, violates RB rule

2) Which of the following arrangement of statues are valid in positions 1,2, and 3 respectively?

Note that in this question, both RB and B3 → Y1 rules must be satisfied. A positioning cannot be correct if the items selected are incorrect.

1	2	3	Valid?
Yellow	Red	Blue	valid, satisfies the rules
Red	Green	Blue	invalid, violates B3 → Y1 rule
Blue	Green	Red	valid, satisfies the rules
Yellow	Blue	Red	valid, satisfies the rules
Green	Red	Yellow	invalid, violates RB rule

Answers

© Gift Of Logic, Inc * Copying prohibited

5   GROUPING AND POSITIONING

Five monkeys m1, m2, m3, m4, and m5..

Cage c1 can hold only a tall monkey.
Cage c3 can hold only a short monkey.
Tall monkeys must be before the short monkeys.

Tall - m1, m2, m3   Short - m4, m5
c1 - one of m1,m2,m3
c3 - one of m4,m5
Tall before short >> m1,m2,m3 before m4,m5

1) Which of the following are valid assignments of monkeys to the three cages?

Cage c1	Cage c2	Cage c3	Valid?
m1	m3	m4	valid
m4	m1	m5	invalid, c1 cannot have short monkey
m1	m5	m3	invalid, c3 cannot have tall monkey
m2	m1	m4	valid

2) If cage c2 can hold only a short monkey, then which of the following can be valid assignments of monkeys to the cages?

Cage c1	Cage c2	Cage c3	Valid?
m2	m3	m4	invalid, c2 has a tall monkey
m2	m4	m5	valid.
m3	m5	m4	valid
m2	m1	m4	invalid, c2 has a tall monkey

Answers
© Gift Of Logic, Inc * Copying prohibited

6 GROUPING AND POSITIONING

Four birds are to be selected from..
Diagram the scenario as follows.

```
group1           group2
b1  b2  b3 | b4  b5  b6
c1  c2  c3  c4
```

\>\>each cage bird from alternate group
\>\> b1,b3 → b1-b3, b2,b6 → b2-b6 (grouping and positioning)

1) Which of the following cage assignments of birds are valid?
Apply the rules to each row to determine if the cages can hold the birds shown.

Cage c1	Cage c2	Cage c3	Cage c4	Valid?
b2	b1	b4	b5	invalid, violates alternate group rule
b3	b5	b1	b6	invalid, violates b1,b3 → b1-b3 rule
b6	b3	b4	b1	invalid, violates b1,b3 → b1-b3 rule
b1	b6	b2	b4	invalid, violates b2,b6 → b2-b6 rule
b6	b2	b5	b3	invalid, violates b2,b6 → b2-b6 rule
b1	b4	b2	b6	valid

2) If birds b1 and b4 are not selected, then which of the following cages cannot hold bird b2?

Answer: D) 4. If b1 and b4 are not selected, then we are left with four birds to fill four cages-namely b2, b3, b5, and b6. Try placing b2 in the four cages and see if it can still satisfy the rules.

Cage c1	Cage c2	Cage c3	Cage c4	Valid?
b2	b5	b3	b6	yes, cage c1 can hold b2
b5	b2	b6	b3	yes, cage c2 can hold b2
b3	b5	b2	b6	yes, cage c3 can hold b2
b5	b3	b6	b2	no, cage c4 cannot hold b2. b2,b6 → b2-b6 rule is violated.

Answers

PATTERN PERCEPTION

Question#	Answer
1	B
2	A
3	A
4	B
5	B
6	B
7	A
8	B

FIGURE FORMATION

Question#	Answer
1	A
2	A
3	B
4	A

PAPER FOLDING AND CUTTING

Question#	Answer
1	D
2	C
3	B

FIGURE MATRIX

Q#	Ans	Reasoning
1	A	circle is 2D, ball is in 3D; square is 2D, dice is in 3D
2	B	small watch, big clock; small printer, big printer
3	A	big dog, small dog; big cat, small cat
4	A	hut is a small house, bungalow is a big house; fish is small, whale is big
5	B	all the three are construction equipment; a cement mixer is also used in construction
6	C	All the three items are used in drawing; an easel board is also used in drawing
7	A	bed, pillow, and dresser belong to the bedroom; mirror also belongs to the bedroom
8	C	all the three are items used in measuring; pencil is used for marking the measurement

RULE DETECTION

Question#	Answer
1	A
2	A
3	B
4	B
5	A
6	B

Answers

NOTES

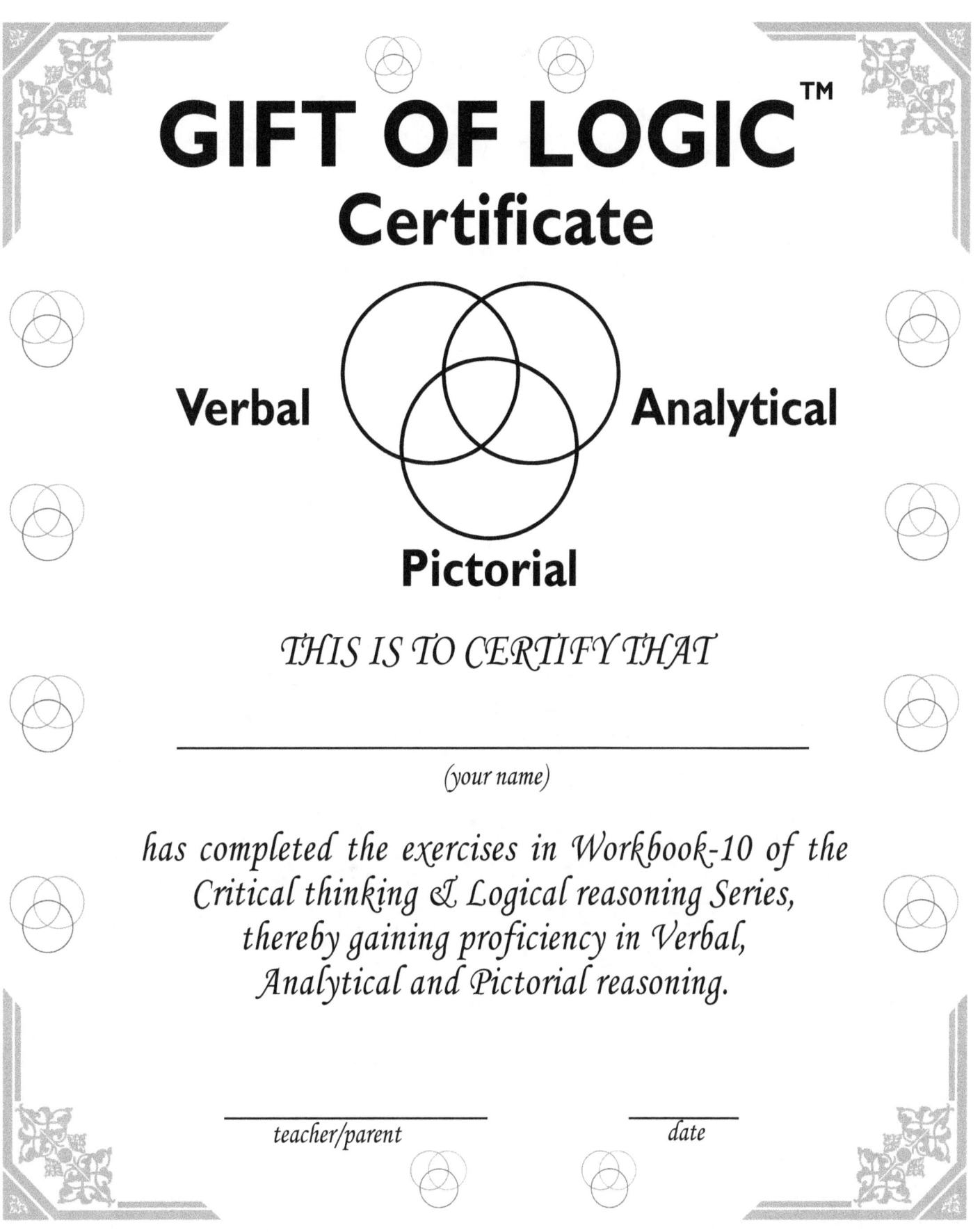

NOTES

www.ingramcontent.com/pod-product-compliance
Lightning Source LLC
Chambersburg PA
CBHW080254180526
45167CB00006B/2527